TEACHER'S EDITION

INVESTIGATING EARTH SYSTEMS™

AN INQUIRY EARTH SCIENCE PROGRAM

INVESTIGATING SOIL

Michael J. Smith Ph.D.
American Geological Institute

John B. Southard Ph.D.
Massachussetts Institute of Technology

Colin Mably
Curriculum Developer

Developed by the American Geological Institute
Supported by the National Science Foundation and
the American Geological Institute Foundation
Endorsed by the Soil Science Society of America

IT'S ABOUT TIME™

Published by
It's About Time Inc., Armonk, NY

It's About Time, Inc.
84 Business Park Drive, Armonk, NY 10504
Phone (914) 273-2233 Fax (914) 273-2227
Toll Free (888) 698-TIME
www.Its-About-Time.com

Publisher
Laurie Kreindler

Project Editor	**Project Coordinator**	**Design**
Ruta Demery	William Jones	John Nordland

Studio Manager	**Associate Editor**
Joan Lee	Al Mari

All student activities in this textbook have been designed to be as safe as possible, and have been reviewed by professionals specifically for that purpose. As well, appropriate warnings concerning potential safety hazards are included where applicable to particular activities. However, responsibility for safety remains with the student, the classroom teacher, the school principals, and the school board.

Investigating Earth Systems™ is a registered trademark of the American Geological Institute. Registered names and trademarks, etc., used in this publication, even without specific indication thereof, are not to be considered unprotected by law.

It's About Time™ is a registered trademark of It's About Time, Inc. Registered names and trademarks, etc., used in this publication, even without specific indication thereof, are not to be considered unprotected by law.

© Copyright 2001: American Geological Institute

All rights reserved. No part of this publication may be reproduced, stored in a retrieval system, or transmitted, in any form or by any means, electronic, mechanical, photocopying, recording, or otherwise, without the prior written permission of the copyright owner.

Care has been taken to trace the ownership of copyright material contained in this publication. The publisher will gladly receive any information that will rectify any reference or credit line in subsequent editions.

Printed and bound in the United States of America

ISBN # 1-58591-086-4

1 2 3 4 5 QC 05 04 03 02 01

This project was supported, in part, by the
National Science Foundation (grant no. 9353035)

Opinions expressed are those of the authors and not necessarily those of the National Science Foundation or the donors of the American Geological Institute Foundation.

Student's Edition Illustrations and Photos

Sx (top right), S15, S39, source U.S. Department of Agriculture

Sx (top left, bottom right), Sxi, S1, S3, S4, S6, S10, S12, S24, S26, S32, S33, S35, S40 (bottom), S42, PhotoDisc

Sx (bottom left), S10, photos by Scott Bauer, Agricultural Research Service, USDA

Sv, Sxi, S2, S7, S16, S20, S27, S44, illustrations by Dennis Falcon

S13, S18, S19, S27, S28, S29, S30, S36, S37, S45, illustrations by Burmar Technical Corporation

S17, courtesy of the Spokane County Soil Survey Team

S23, S24, map source U.S. Geological Survey

S31, S46, photos by Eric Shih, American Geological Institute

S12, source G. Mathieson Federal Emergency Management Agency News Photo

S40 (top), photo by Jack Dykinga, Agricultural Research Service, USDA

Taking Full Advantage of Investigating Earth Systems Through Professional Development

Implementing a new curriculum is challenging. That is why It's About Time Publishing has partnered with the American Geological Institute, developers of *Investigating Earth Systems (IES)*, to provide a full range of professional development services. The sessions described below were designed to help you deepen your understanding of the content, pedagogy, and assessment strategies outlined in this Teacher's Edition, and adapt the program to suit the needs of your students and your local and state standards and curriculum frameworks.

Professional Development Services Available

Implementation Workshops
Two to five-day sessions held at your site that prepare you to implement the inquiry, systems, and community-based approach to learning Earth Science featured in *IES*. These workshops can be tailored to serve the needs of your school district, with chapters selected from the modules based on local or state curricula and framework criteria.

Program Overviews
One to three-day introductory sessions that provide a complete overview of the content and pedagogy of the *IES* program, as well as hands-on experience with activities from specific chapters. Program overviews are designed in consultation with school districts, counties, and SSI organizations.

Regional New-Teacher Summer Institutes
Two to five-day sessions that are designed to deepen your Earth Science content knowledge, and to prepare you to teach through inquiry. Guidance is provided in the gathering and use of appropriate materials and resources and specific attention is directed to the assessment of student learning.

Leadership Institutes
Six-day summer sessions conducted by the American Geological Institute that are designed to prepare current users for professional development leadership and mentoring within their districts or as consultants for It's About Time.

Follow-up Workshops
One to two-day sessions that provide additional Earth Science content and pedagogy support to teachers using the program. These workshops focus on identifying and solving practical issues and challenges to implementing an inquiry-based program.

Mentoring Visits
One-day visits that can be tailored to your specific needs that include class visits, mentoring teachers of the program, and in-service sessions.

Please fill in the form below to receive more information about participating in one of these Professional Development Services. The form can be directly faxed to our Professional Development at 914-273-2227. Our department will contact you to discuss further details and fees.

District/School: _____ Phone: _____

Address: _____

Contact Name: _____ Title: _____

E-mail: _____ Fax: _____

School Enrollment: _____ Number of Students Impacted: _____ Grade Level: _____

Have you purchased the following: ❏ Student Editions   ❏ Teacher Editions   ❏ Kits

Briefly explain how you plan to implement or how you are implementing the program in your school.

_____

_____

Teacher's Edition

Table of Contents

Investigating Earth Systems Team	vi
Acknowledgements	viii
The American Geological Institute and *Investigating Earth Systems*	xi
Developing *Investigating Earth Systems*	xii
Investigating Earth Systems Modules	xiii
Investigating Earth Systems: Correlation to the National Science Education Standards	xiv
Using *Investigating Earth Systems* Features in Your Classroom	xvi
Using the *Investigating Earth Systems* Web Site	xxx
Enhancing Teacher Content Knowledge	xxxi
Managing Inquiry in Your *Investigating Earth Systems* Classroom	xxxii
Assessing Student Learning in *Investigating Earth Systems*	xxxv
Investigating Earth Systems Assessment Tools	xxxviii
Reviewing and Reflecting upon Your Teaching	xlii
Investigating Soil: Introduction	1
Students' Conceptions about Soil	3
Investigating Soil: Module Flow	5
Investigating Soil: Module Objectives	6
National Science Education Content Standards	9
Key NSES Earth Science Standards Addressed in *IES* Soil	10
Key AAAS Earth Science Benchmarks Addressed in *IES* Soil	10
Materials and Equipment List for Investigating Soil	11
Pre-assessment	14
Introducing the Earth System	19
Introducing Inquiry Processes	21
Introducing Soil	23
Why is Soil Important?	25
Investigation 1: Beginning to Investigate Soil	27
Investigation 2: Separating Soil by Settling	57
Investigation 3: Separating Soil by Sieving	87
Investigation 4: Examining Core Samples of Soil	111
Investigation 5: Water and Other Chemicals in Soil	147
Investigation 6: Soil Erosion	185
Investigation 7: Using Soil Data to Plan a Garden	215
Reflecting	235
Appendices: Alternative End-of-Module Assessment	238
Assessment Tools	242
Blackline Masters	252

Investigating Earth Systems Team

Project Staff

Michael J. Smith, Principal Investigator
  Director of Education, American Geological Institute
John B. Southard, Senior Writer
  Professor of Geology, Massachusetts Institute of Technology
William O. Jones, Project Assistant
  American Geological Institute
Caitlin N. Callahan, Project Assistant
  American Geological Institute
William S. Houston, Field Test Coordinator
  American Geological Institute
Harvey Rosenbaum, Field Test Evaluator
  Montgomery County School District, Maryland
Fred Finley, Project Evaluator
  University of Minnesota
Lynn Lindow, Pilot Test Evaluator
  University of Minnesota

Original Project Personnel

Robert L. Heller, Principal Investigator
Charles Groat, United States Geological Survey
Colin Mably, LaPlata, Maryland
Robert Ridky, University of Maryland
Marilyn Suiter, American Geological Institute

Teacher's Edition

National Advisory Board

Jane Crowder
   Middle School Teacher, WA
Kerry Davidson
   Louisiana Board of Regents, LA
Joseph D. Exline
   Educational Consultant, VA
Louis A. Fernandez
   California State University, CA
Frank Watt Ireton
   National Earth Science Teachers Association, DC
LeRoy Lee
   Wisconsin Academy of Sciences, Arts and Letters, WI
Donald W. Lewis
   Chevron Corporation, CA
James V. O'Connor (deceased)
   University of the District of Columbia, DC
Roger A. Pielke Sr.
   Colorado State University, CO
Dorothy Stout
   Cypress College, CA
Lois Veath
   Advisory Board Chairperson - Chadron State College, NE

National Science Foundation Program Officers

Gerhard Salinger
Patricia Morse

Acknowledgements

Principal Investigator

Michael Smith is Director of Education at the American Geological Institute in Alexandria, Virginia. Dr. Smith worked as an exploration geologist and hydrogeologist. He began his Earth Science teaching career with Shady Side Academy in Pittsburgh, PA in 1988 and most recently taught Earth Science at the Charter School of Wilmington, DE. He earned a doctorate from the University of Pittsburgh's Cognitive Studies in Education Program and joined the faculty of the University of Delaware School of Education in 1995. Dr. Smith received the Outstanding Earth Science Teacher Award for Pennsylvania from the National Association of Geoscience Teachers in 1991, served as Secretary of the National Earth Science Teachers Association, and is a reviewer for Science Education and The Journal of Research in Science Teaching. He worked on the Delaware Teacher Standards, Delaware Science Assessment, National Board of Teacher Certification, and AAAS Project 2061 Curriculum Evaluation programs.

Senior Writer

John Southard received his undergraduate degree from the Massachusetts Institute of Technology in 1960 and his doctorate in geology from Harvard University in 1966. After a National Science Foundation postdoctoral fellowship at the California Institute of Technology, he joined the faculty at the Massachusetts Institute of Technology, where he is currently Professor of Geology. He was awarded the MIT School of Science teaching prize in 1989 and was one of the first cohorts of first MacVicar Fellows at MIT, in recognition of excellence in undergraduate teaching. He has taught numerous undergraduate courses in introductory geology, sedimentary geology, field geology, and environmental Earth Science both at MIT and in Harvard's adult education program. He was editor of the Journal of Sedimentary Petrology from 1992 to 1996, and he continues to do technical editing of scientific books and papers for SEPM, a professional society for sedimentary geology.

Project Director/Curriculum Designer

Colin Mably has been a key curriculum developer for several NSF-supported national curriculum projects. As learning materials designer to the American Geological Institute, he has directed the design and development of the IES curriculum modules and also training workshops for pilot and field-test teachers.

Teacher's Edition

Project Team

Marcus Milling
Executive Director - AGI, VA

Michael Smith
Principal Investigator
Director of Education - AGI, VA

Colin Mably
Project Director/Curriculum Designer
Educational Visions, MD

Fred Finley
Project Evaluator
University of Minnesota, MN

Lynn Lindow
Pilot Test Evaluator
University of Minnesota, MN

Harvey Rosenbaum
Field Test Evaluator
Montgomery School District, MD

Ann Benbow
Project Advisor - American Chemical Society, DC

Robert Ridky
Original Project Director
University of Maryland, MD

Chip Groat
Original Principal Investigator
University of Texas - El Paso, TX

Marilyn Suiter
Original Co-principal Investigator
AGI, VA

William Houston
Project Manager

Eric Shih - Project Assistant

Original and Contributing Authors

Oceans
George Dawson
Florida State University, FL
Joseph F. Donoghue
Florida State University, FL
Ann Benbow
American Chemical Society
Michael Smith
American Geological Institute

Soil
Robert Ridky
University of Maryland, MD
Colin Mably - LaPlata, MD
John Southard
Massachusetts Institute of Technology, MA
Michael Smith
American Geological Institute

Fossils
Robert Gastaldo
Colby College, ME
Colin Mably - LaPlata, MD
Michael Smith
American Geological Institute

Climate and Weather
Mike Mogil
How the Weather Works, MD

Ann Benbow
American Chemical Society
Michael Smith
American Geological Institute

Energy Resources
Laurie Martin-Vermilyea
American Geological Institute
Michael Smith
American Geological Institute

Dynamic Planet
Michael Smith
American Geological Institute

Rocks and Landforms
Michael Smith
American Geological Institute

Water as a Resource
Ann Benbow
American Chemical Society
Michael Smith
American Geological Institute

Materials and Minerals
Mary Poulton
University of Arizona, AZ
Colin Mably - LaPlata, MD
Michael Smith
American Geological Institute

Content Reviewers

Louis Bartek
University of North Carolina
Gary Beck - BP Exploration
Steve Bergman
University of Texas-Dallas
Joseph Bishop
Johns Hopkins University/NOAA
Kathleen Carrado
Argonne National Laboratory
Sandip Chattopadhyay
R.S. Kerr Environmental Research Center
Bob Christman
Western Washington University
Donald Conte
California University of California
Norbert E. Cygan - AAPG
Tom Dignes
Mobil Technology Corporation
Neil M. Dubrovsky
United States Geological Survey
Robert J. Finley
Illinois State Geological Survey
Anke Friedrich
California Institute of Technology
Rick Fritz - AAPG
Frank Hall - University of New Orleans
David Hawkins - Denison University
Martha House
California Institute of Technology
Travis Hudson
American Geological Institute
Allan P. Juhas - SEG
Dennis Lamb - Penn State
Donald Lewis - Happy Valley, CA

Kate Madin
Woods Hole Oceanographic Institute
John Madsen - University of Delaware
Carol Mankiewicz - Beloit College
Clyde J. Northrup
Boise State University
Lois K. Ongley, PhD - Bates College
Bruce Pivetz
ManTech Environmental Research Services
Eleanora I. Robbins
United States Geological Survey
Rob Ross
Paleontological Research Institution
Audrey Rule - Boise State University
Lou Solebello - Macon, GA
Steve Stanley - Johns Hopkins University
Sarah Tebbens - University of South Florida
Bob Tilling
United States Geological Survey
Michael Velbel
Michigan State University
Don Woodrow
Hobart and William Smith Colleges

Pilot Test Teachers

Debbie Bambino - Philadelphia, PA
Barbara Barden - Rittman, OH
Louisa Bliss - Bethlehem, NH
Mike Bradshaw - Houston TX
Greta Branch - Reno, NV
Garnetta Chain - Piscataway, NJ
Roy Chambers - Portland, OR
James Cole - New York, NY
Laurie Corbett - Sayre, PA
Collette Craig - Reno, NV
Anne Douglas - Houston, TX
Jacqueline Dubin - Roslyn, PA
Jane Evans - Media, PA
Gail Gant - Houston, TX
Joan Gentry - Houston, TX
Pat Gram - Aurora, OH
Robert Haffner - Akron, OH
Joe Hampel - Swarthmore, PA
Wayne Hayes - West Green, GA
Mark Johnson - Reno, NV
Cheryl Joloza - Philadelphia, PA
Jeff Luckey - Houston, TX
Karen Luniewski
Reistertown, MD
Cassie Major - Plainfield, VT
Carol Miller - Houston, TX
Melissa Murray - Reno, NV
Mary-Lou Northrop
North Kingstown, RI
Keith Olive - Ellensburg, WA

Investigating Earth Systems: Soil

Tracey Oliver - Philadelphia, PA
Nicole Pfister - Londonderry, VT
Beth Price - Reno, NV
Joyce Ramig - Houston, TX
Julie Revilla - Woodbridge, VA
Steve Roberts - Meredith, NH
Cheryl Skipworth - Philadelphia, PA
Brent Stenson - Valdosta, GA
Elva Stout - Evans, GA
Regina Toscani - Philadelphia, PA
Bill Waterhouse - North Woodstock, NH
Leonard White - Philadelphia, PA
Paul Williams - Lowerford, VT
Bob Zafran - San Jose, CA
Missi Zender - Twinsburg, OH

Field Test Teachers
Eric Anderson - Carson City, NV
Katie Bauer - Rockport, ME
Kathleen Berdel - Philadelphia, PA
Wanda Blake - Macon, GA
Beverly Bowers - Mannington, WV
Rick Chiera - Monroe Falls, OH
Don Cole - Akron, OH
Patte Cotner - Bossier City, LA
Johnny DeFreese - Haughton, LA
Mary Devine - Astoria, NY
Cheryl Dodes - Queens, NY
Brenda Engstrom - Warwick, RI
Lisa Gioe-Cordi - Brooklyn, NY
Pat Gram - Aurora, OH
Mark Johnson - Reno, NV
Chicory Koren - Kent, OH
Marilyn Krupnick - Philadelphia, PA

Melissa Loftin - Bossier City, LA
Janet Lundy - Reno, NV
Vaughn Martin - Easton, ME
Anita Mathis - Fort Valley, GA
Laurie Newton - Truckee, NV
Debbie O'Gorman - Reno, NV
Joe Parlier - Barnesville, GA
Sunny Posey - Bossier City, LA
Beth Price - Reno, NV
Stan Robinson - Mannington, WV
Mandy Thorne - Mannington, WV
Marti Tomko - Westminster, MD
Jim Trogden - Rittman, OH
Torri Weed - Stonington, ME
Gene Winegart - Shreveport, LA
Dawn Wise - Peru, ME
Paula Wright - Gray, GA

IMPORTANT NOTICE

The *Investigating Earth Systems*™ series of modules is intended for use by students under the direct supervision of a qualified teacher. The experiments described in this book involve substances that may be harmful if they are misused or if the procedures described are not followed. Read cautions carefully and follow all directions. Do not use or combine any substances or materials not specifically called for in carrying out experiments. Other substances are mentioned for educational purposes only and should not be used by students unless the instructions specifically indicate.

The materials, safety information, and procedures contained in this book are believed to be reliable. This information and these procedures should serve only as a starting point for classroom or laboratory practices, and they do not purport to specify minimal legal standards or to represent the policy of the American Geological Institute. No warranty, guarantee, or representation is made by the American Geological Institute as to the accuracy or specificity of the information contained herein, and the American Geological Institute assumes no responsibility in connection therewith. The added safety information is intended to provide basic guidelines for safe practices. It cannot be assumed that all necessary warnings and precautionary measures are contained in the printed material and that other additional information and measures may not be required.

This work is based upon work supported by the National Science Foundation under Grant No. 9353035 with additional support from the Chevron Corporation. Any opinions, findings, and conclusions or recommendations expressed in this publication are those of the authors and do not necessarily reflect the views of the National Science Foundation or the Chevron Corporation. Any mention of trade names does not imply endorsement from the National Science Foundation or the Chevron Corporation.

Teacher's Edition

The American Geological Institute and Investigating Earth Systems

Imagine more than 500,000 Earth scientists worldwide sharing a common voice, and you've just imagined the mission of the American Geological Institute. Our mission is to raise public awareness of the Earth sciences and the role that they play in mankind's use of natural resources, mitigation of natural hazards, and stewardship of the environment. For more than 50 years, AGI has served the scientists and teachers of its Member Societies and hundreds of associated colleges, universities, and corporations by producing Earth science educational materials, *Geotimes*–a geoscience news magazine, GeoRef–a reference database, and government affairs and public awareness programs.

So many important decisions made every day that affect our lives depend upon an understanding of how our Earth works. That's why AGI created *Investigating Earth Systems*. In your *Investigating Earth Systems* classroom, you'll discover the wonder and importance of Earth science. As you investigate minerals, soil, or oceans — do field work in nearby beaches, parks, or streams, explore how fossils form, understand where your energy resources come from, or find out how to forecast weather — you'll gain a better understanding of Earth science and its importance in your life.

We would like to thank the National Science Foundation and the AGI Foundation Members that have been supportive in bringing Earth science to students. The Chevron Corporation provided the initial leadership grant, with additional contributions from the following AGI Foundation Members: Anadarko Petroleum Corp., Baker Hughes Foundation, Barrett Resources Corp., BPAmoco Foundation, Burlington Resources Foundation, Conoco Inc., Consolidated Natural Gas Foundation, Diamond Offshore Co., EEX Corp., ExxonMobil Foundation, Global Marine Drilling Co., Halliburton Foundation, Inc., Kerr McGee Foundation, Maxus Energy Corp., Noble Drilling Corp., Occidental Petroleum Charitable Foundation, Parker Drilling Co., Phillips Petroleum Co., Santa Fe Snyder Corp., Schlumberger Foundation, Shell Oil Company Foundation, Southwestern Energy Co., Texaco, Inc., Texas Crude Energy, Inc., Unocal Corp. USX Foundation (Marathon Oil Co.).

We at AGI wish you success in your exploration of the Earth System!

Michael J. Smith
Director of Education, AGI

Marcus E. Milling
Executive Director, AGI

Developing Investigating Earth Systems

Welcome to *Investigating Earth Systems (IES)! IES* was developed through funding from the National Science Foundation and the American Geological Institute Foundation. Classroom teachers, scientists, and thousands of students across America helped to create *IES*. In the 1997-98 school year, scientists and curriculum developers drafted nine *IES* modules. They were pilot tested by 43 teachers in 14 states from Washington to Georgia. Faculty from the University of Minnesota conducted an independent evaluation of the pilot test in 1998, which was used to revise the program for a nationwide field test during the 1999-2000 school year. A comprehensive evaluation of student learning by a professional field-test evaluator showed that *IES* modules led to significant gains in student understanding of fundamental Earth science concepts. Field-test feedback from 34 teachers and content reviews from 33 professional Earth scientists were used to produce the commercial edition you have selected for your classroom.

Inquiry and the interrelation of Earth's systems form the backbone of *IES*. Often taught as a linear sequence of events called "the scientific method," inquiry underlies all scientific processes and can take many different forms. It is very important that students develop an understanding of inquiry processes as they use them. Your students naturally use inquiry processes when they solve problems. Like scientists, students usually form a question to investigate after first looking at what is observable or known. They predict the most likely answer to a question. They base this prediction on what they already know to be true. Unlike professional scientists, your students may not devote much thought to these processes. In order to be objective, students must formally recognize these processes as they do them. To make sure that the way they test ideas is fair, scientists think very carefully about the design of their investigations. This is a skill your students will practice throughout each *IES* module.

All *Investigating Earth Systems* modules also encourage students to think about the Earth as a system. Upon completing each investigation they are asked to relate what they have learned to the Earth Systems (see the *Earth System Connection* sheet in the **Appendix**). Integrating the processes of the biosphere, geosphere, hydrosphere, and atmosphere will open up a new way of looking at the world for most students. Understanding that the Earth is dynamic and that it affects living things, often in unexpected ways, will engage them and make the topics more relevant.

We trust that you will find the Teacher's Edition that accompanies each student module to be useful. It provides **Background Information** on the concepts explored in the module, as well as strategies for incorporating inquiry and a systems-based approach into your classroom. Enjoy your investigation!

Teacher's Edition

Investigating Earth Systems Modules

Climate and Weather

Dynamic Planet

Energy Resources

Fossils

Materials and Minerals

Oceans

Rocks and Landforms

Soil

Water as a Resource

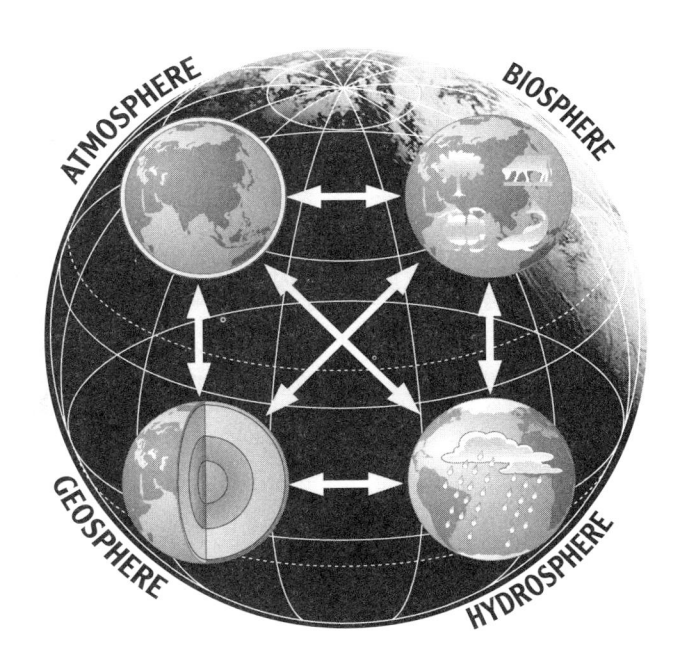

Investigating Earth Systems: Correlation to the National Science Education Standards

National Science Education Content Standards Grades 5 – 8	Soil	Rocks and Landforms	Oceans	Climate and Weather	Dynamic Planet	Materials and Minerals	Energy Resources	Water as a Resource	Fossils
UNIFYING CONCEPTS AND PROCESSES									
System, order and organization	•	•	•	•	•	•	•	•	•
Evidence, models, and explanation	•	•	•	•	•	•	•	•	•
Constancy, change, and measurement	•	•	•	•	•	•	•	•	•
Evolution and equilibrium		•	•	•	•			•	•
Form and function									•
SCIENCE AS INQUIRY									
Identify questions that can be answered through scientific investigations	•	•	•	•	•	•	•	•	•
Design and conduct scientific investigations	•	•		•	•	•	•	•	•
Use tools and techniques to gather, analyze, and interpret data	•	•	•	•	•	•	•	•	•
Develop descriptions, explanations, predictions and models based on evidence	•	•		•	•	•	•	•	•
Think critically and logically to make the relationships between evidence and explanation	•	•	•	•	•	•	•	•	•
Recognize and analyze alternative explanations and predictions	•	•	•	•	•	•	•	•	•
Communicate scientific procedures and explanations	•	•	•	•	•	•	•	•	•
Use mathematics in all aspects of scientific inquiry	•	•	•	•	•	•	•	•	•
Understand scientific inquiry	•	•	•	•	•	•	•	•	•
PHYSICAL SCIENCE									
Properties and Changes of Properties in Matter	•	•	•		•	•	•	•	
Motions and Forces	•		•						
Transfer of Energy		•	•	•	•	•	•	•	
LIFE SCIENCE									
Populations and Ecosystems			•				•	•	•
Diversity and Adaptation of Organisms			•		•				•

Teacher's Edition

Investigating Earth Systems: Correlation to the National Science Education Standards

National Science Education Content Standards Grades 5 – 8

	Soil	Rocks and Landforms	Oceans	Climate and Weather	Dynamic Planet	Materials and Minerals	Energy Resources	Water as a Resource	Fossils
EARTH AND SPACE SCIENCE									
Structure of the Earth system	•	•	•	•	•	•	•	•	•
Earth's History	•	•	•	•	•	•	•	•	•
Earth in the Solar System			•	•	•		•	•	
SCIENCE AND TECHNOLOGY									
Abilities of technological design	•	•	•	•	•	•	•	•	•
Understandings about science and technology		•	•			•	•	•	
SCIENCE IN PERSONAL AND SOCIAL PERSPECTIVES									
Personal health	•							•	
Populations, resources, and environment	•					•	•	•	
Natural Hazards		•		•	•	•			
Risks and benefits				•			•		
Science and technology in society	•	•	•	•	•	•	•	•	•
HISTORY AND NATURE OF SCIENCE									
Science as a human endeavor	•	•	•	•	•	•	•	•	•
Nature of science	•	•	•	•	•	•	•	•	•
History of science			•		•				•

Investigating Earth Systems: Soil

Using Investigating Earth Systems Features in Your Classroom

1. Pre-assessment

Designed under the umbrella framework of "science for all students," meaning that all students should be able to engage in inquiry and learn core science concepts, *Investigating Earth Systems* helps you to tailor instruction to meet your students' needs. A crucial first step in this framework is to ascertain what knowledge, experience, and understanding your students bring to their study of a module. The pre-assessment consists of four questions geared to the major concepts and understandings targeted in the unit. Students write and draw what they know about the major topics and concepts. This information is recorded and shared in an informal discussion prior to engaging in hands-on inquiry. The discussion enables students to recognize how much there is to learn and appreciate, and that by exploring the unit together, the entire classroom can emerge from the experience with a better understanding of core concepts and themes. Students' responses provide crucial pre-assessment data for you. By examining their written work and probing for further detail during the classroom conversation, you can identify strengths and weaknesses in students' understandings, as well as their abilities to communicate that understanding to others. It is important that the pre-assessment not be viewed as a test, and that judgments about the accuracy of responses be evaluated in writing or through your comments during the conversation. The goal is to ascertain and probe, not judge, and to create a safe classroom environment in which students feel comfortable sharing their ideas and knowledge. Students revisit these pre-assessment questions informally throughout the unit. At the end of the unit, students respond to the same four questions in the section called **Back to the Beginning**. The pre-assessment thus helps you and your students to make judgments about their growth in understanding and ability throughout the module.

Teacher's Edition

2. The Earth System

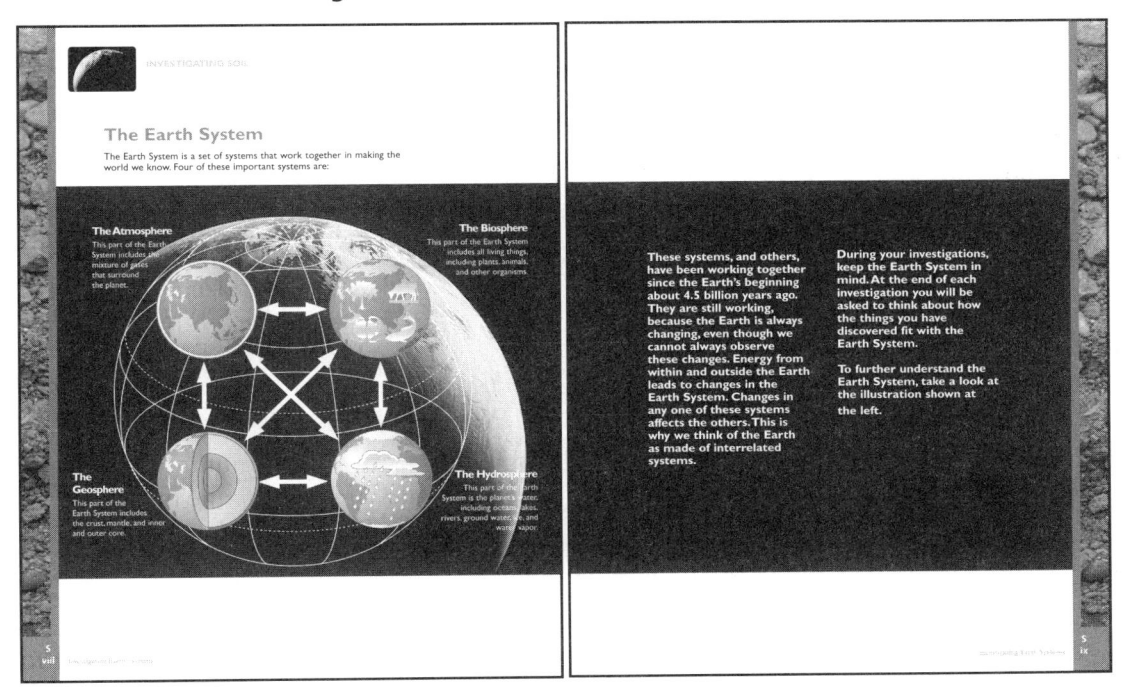

National Science Education Standards link...

"A major goal of science in the middle grades is for students to develop an understanding of Earth (and the solar system) as a set of closely coupled systems. The idea of systems provides a framework in which students can investigate the four major interacting components of the Earth System – geosphere (crust, mantle, and core), hydrosphere (water), atmosphere (air), and the biosphere (the realm of living things)."

NSES content standard D "Developing Student Understanding" (pages 158-159)

Understanding the Earth system is an overall goal of the *Investigating Earth Systems* series. It is a difficult and complex set of concepts to grasp, because it is inferred rather than observed directly. Yet even the smallest component of Earth science can be linked to the Earth system. As your students progress through each module, an increasing number of connections with the Earth system will arise. Your students may not, however, immediately see these connections. At the end of every investigation, they will be asked to link what they have discovered with ideas about the Earth system. They will also be asked to write about this in their journals. A **Blackline Master** (*Earth System Connection* sheet) is available in each Teacher's Edition. Students can use this to record connections that they make as they complete each investigation. At the very end of the module they will be asked to review everything they have learned in relation to the Earth system. The aim is for students to have a working understanding of the Earth System by the time they complete grade 8. They will need your help accomplishing this.

Investigating Earth Systems: Soil

For example, in *Investigating Rocks and Landforms*, students work with models to simulate Earth processes, such as erosion of stream sediment and deposition of that sediment on floodplains and in deltas. Changes in inputs in one part of the system (say rainfall, from the atmosphere), affect other parts of the system (stream flows, erosion on river bends, amount of sediment carried by the stream, and deposition of sediment on floodplains or in deltas). These changes affect, in turn, other parts of the system (for example, floods that affect human populations, i.e., the biosphere). In the same module, students explore the rock record within their community and develop understandings about how interactions between the hydrosphere, atmosphere, geosphere, and biosphere change the landscape over time. These are just some of the many ways that *Investigating Earth Systems* modules foster and promote student thinking about the dynamic nature and interactions of Earth systems—biosphere, geosphere, atmosphere, and hydrosphere.

3. Introducing Inquiry Processes

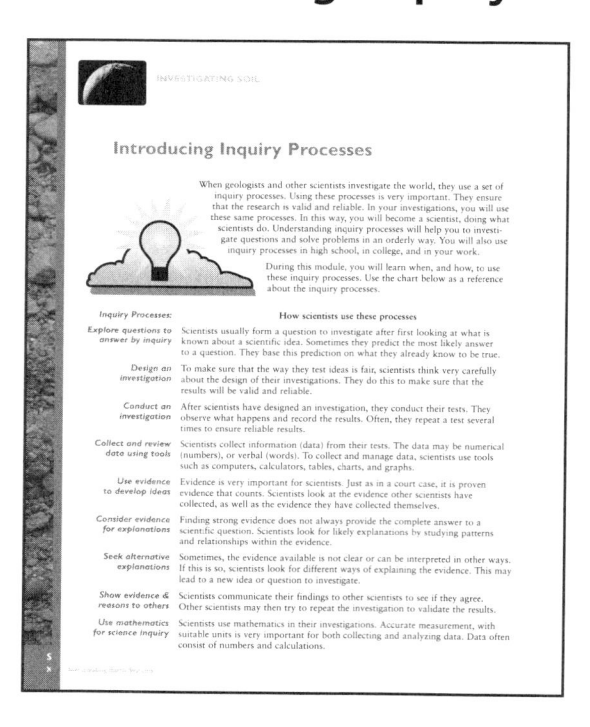

Inquiry is at the heart of *Investigating Earth Systems*. That is why each module title begins with the title "Investigating." In the National Science Education Standards, inquiry is the first Content Standard. NSES then lists a range of points about inquiry. These fundamental components of inquiry were written into the list shown at the beginning of each student module. It is very important that students be reminded of the steps in the inquiry process as they perform them. Inquiry depends on active student participation. Ideas on how to make inquiry successful in the classroom appear throughout the modules and in the "Managing Inquiry in Your *Investigating Earth Systems* Classroom" section of this Teacher's Edition.

It is very important that students develop an understanding of the inquiry processes as they use them. Stress the importance of inquiry processes as they occur in your investigations. Provoke students to think about why these processes are important. Collecting good data, using evidence, considering alternative explanations, showing evidence to others, and using mathematics are all essential to *IES*. Use examples to demonstrate these processes whenever possible. At the end of every investigation, students are asked to reflect on the scientific inquiry processes they used. Refer students to the list of inquiry processes on page x of the Student Book as they think about scientific inquiry and answer the questions.

4. Introducing the Module

Each *IES* module begins with photographs and questions. This is an introduction to the module for your students. It is designed to give them a brief overview of the content of the module and set their investigations into a relevant and meaningful context. Students will have had a variety of experiences with the content of the module. This is an opportunity for them to offer some of their own experiences in a general discussion, using these questions as prompts. This section of each *IES* module follows the pre-assessment, where students spend time thinking about what they already know about the content of the module. The photographs and questions can be used to focus the students' thinking.

The ideas students share in the introduction to the module provide you with additional pre-assessment data. The experiences they describe and the way in which they are discussed will alert you to their general level of understanding about these topics. To encourage sharing and to provide a record, teachers find it useful to quickly summarize the main points that emerge from discussion. You can do this on the chalkboard or flipchart for all to see. This can be displayed as students work through the module and added to with each new experience. For your own assessment purposes, it will be useful to keep a record of these early indicators of student understanding.

5. Key Question

Each *Investigating Earth Systems* investigation begins with a **Key Question** – an open-ended question that gives teachers the opportunity to explore what their students know about the central concepts of the activity. Uncovering students' thinking (their prior knowledge) and exposing the diversity of ideas in the classroom are the first steps in the learning cycle. One of the most fundamental principles derived from many years of research on student learning is that:

"Students come to the classroom with preconceptions about how the world works. If their initial understanding is not engaged, they may fail to grasp the new concepts and information that are taught, or they may learn them for the purposes of a test but revert to their preconceptions outside the classroom." (*How People Learn: Bridging Research and Practice*, National Research Council, 1999, P. 10.)

This principle has been illustrated through the *Private Universe* series of videotapes that show Harvard graduates responding to basic science questions in much the same way that fourth grade students do. Although the videotapes revealed that the Harvard graduates used a more sophisticated vocabulary, the majority held onto the same naïve, incorrect conceptions of elementary school students. Research on learning suggests that the belief systems of students who are not confronted with what they believe and adequately shown why they should give up that belief system remain intact. Real learning requires confronting one's beliefs and testing them in light of competing explanations.

Drawing out and working with students' preconceptions is important for learners. In *Investigating Earth Systems*, the **Key Question** is used to ascertain students' prior knowledge about the key concept or Earth science processes or events explored in the activity. Students verbalize what they think about the age of the Earth, the causes of volcanoes, or the way that the landscape changes over time before they embark on an activity designed to challenge and test these beliefs. A brief discussion about the diversity of beliefs in the classroom makes students consider how their ideas compare to others and the evidence that supports their view of volcanoes, earthquakes, or seasons.

Teacher's Edition

The **Key Question** is not a conclusion, but a lead into inquiry. It is not designed to instantly yield the "correct answer" or a debate about the features of the question, or to bring closure. The activity that follows will provide that discussion as students analyze and discuss the results of inquiry. Students are encouraged to record their ideas in words and/or drawings to ensure that they have considered their prior knowledge. After students discuss their ideas in pairs or in small groups, teachers activate a class discussion. A discussion with fellow students prior to class discussion may encourage students to exchange ideas without the fear of personally giving a "wrong answer." Teachers sometimes have students exchange papers and volunteer responses that they find interesting.

Some teachers prefer to have students record their responses to these questions. They then call for volunteers to offer ideas up for discussion. Other teachers prefer to start with discussion by asking students to volunteer their ideas. In either situation, it is important that teachers encourage the sharing of ideas by not judging responses as "right" or "wrong." It is also important that teachers keep a record of the variety of ideas, which can be displayed in the classroom (on a sheet of easel pad paper or on an overhead transparency) and referred to as students explore the concepts in the module. Teachers often find that they can group responses into a few categories and record the number of students who hold each idea. The photograph in each **Key Question** section was designed to stimulate student thinking and help students to make the specific kinds of connections emphasized in each activity.

6. Investigate

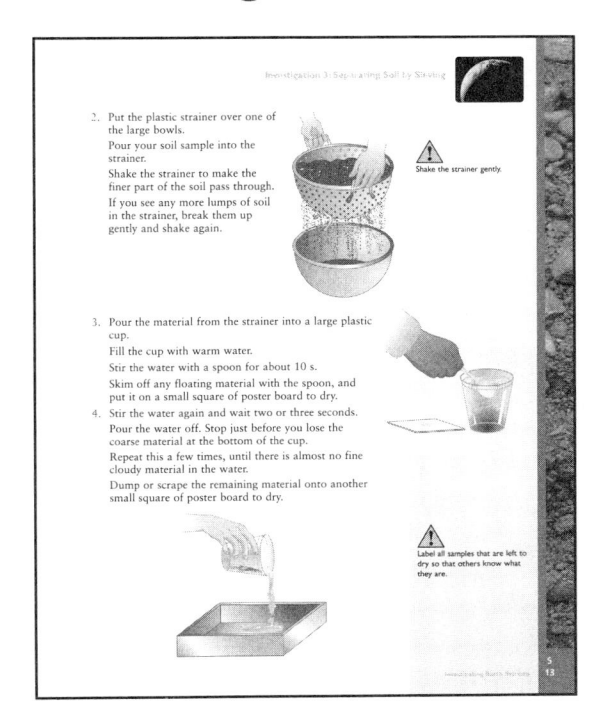

Investigating Earth Systems is a hands-on, minds-on curriculum. In designing *Investigating Earth Systems*, we were guided by the research on learning, which points out how important *doing* Earth Science is to *learning* Earth Science. Testing of *Investigating Earth Systems* activities by teachers across America provided critical testimonial and quantitative measures of the importance of the activities to student learning. In small groups and as a class, students take part in doing hands-on experiments, participating in field work, or searching for answers using the Internet and reference materials. **Blackline Masters** are included in the Teacher's Editions for any maps or illustrations that are essential for students to complete the activity.

Each part of an *Investigating Earth Systems* investigation, as well as the sequence of activities within a module, moves from concrete to abstract. Hands-on activities provide the basis for exploring student beliefs about how the world works and to manipulate variables that affect the outcomes of experiments, models, or simulations. Later in each activity, formal labels are applied to concepts by introducing terminology used to describe the processes that students have explored through hands-on activity. This flow from concrete (hands-on) to abstract (formal explanations) is progressive — students begin to develop their own explanations for phenomena by responding to questions within the **Investigate** section.

Each activity has instructions for each part of the investigation. Materials kits are available for purchase, but you will also need to obtain some resources from outside suppliers, such as topographic and geologic maps of your community, state, or region. The *Investigating Earth Systems* web site will direct you to sources where you can gather such materials.

Most **Investigate** activities will require between one and two class periods. The variety of school schedules and student needs makes it difficult to predict exactly how much time your class will need. For example, if students need to construct a graph for part of an investigation, and the students have never been exposed to graphing, then this investigation may require additional time and could become part of a mathematics lesson.

The most challenging aspect of *Investigating Earth Systems* for teachers to "master" is that the **Investigate** section of each activity has been designed to be student-driven. Students learn more when they have to struggle to "figure things out" and work in collaborative groups to solve problems as a team. Teachers will have to resist the temptation to provide the answers to students when they get "stuck" or hung up on part of a problem. Eventually, students learn that while they can call upon their teacher for assistance, the teacher is not going to "show them the answer." Field testing of *Investigating Earth Systems* revealed that teachers who were most successful in getting their students to solve problems as a team were patient with this process and steadfast in their determination to act as facilitators of learning during the **Investigate** portion of activities. As one teacher noted, "My response to questions during the investigation was like a mantra, 'What do you think you need to do to solve this?' My students eventually realized that although I was there to provide guidance, they weren't going to get the solution out of me."

Another concern that many teachers have when examining *Investigating Earth Systems* for the first time is that their students do not have the background knowledge to do the investigations. They want to deliver a lecture about the phenomena before allowing students to do the investigation. Such an approach is common to many traditional programs and is inconsistent with the pedagogical theory used to design *Investigating Earth Systems*. The appropriate place for delivering a lecture or reading text in *Investigating Earth Systems* is following the investigation, not preceding it.

Teacher's Edition

For example, suppose a group of students has been asked to interpret a map. The traditional approach to science education is for the teacher to give a lecture or assign a reading, "How to Interpret Maps," then give students practice reading maps. *Investigating Earth Systems* teachers recognize that while students may lack some specific skills (reading latitude and longitude, for example), within a group of four students, it is not uncommon for at least one of the students to have a vital skill or piece of knowledge that is required to solve a problem. The one or two students who have been exposed to (or better yet, understood) latitude and longitude have the opportunity to shine within the group by contributing that vital piece of information or demonstrating a skill. That's how scientific research teams work – specialists bring expertise to the group, and by working together, the group achieves something that no one could achieve working alone. The **Investigate** section of *Investigating Earth Systems* is modeled in the spirit of the scientific research team.

7. Inquiry

Inquiry is the first content standard in the National Science Education Standards (NSES). The American Association for the Advancement of Science's (AAAS) Benchmarks for Science Literacy also places considerable emphasis on scientific inquiry (see excerpts on the following page). *IES* has been designed to remind students to reflect on inquiry processes as they carry out their investigations. The student journal is an important tool in helping students to develop these understandings. In using the journal, students are modeling what scientists do. Your students are young scientists as they investigate Earth science questions. Encourage your students to think of themselves in this way and to see their journals as records of their investigations.

Inquiry
Representing Information

Communicating findings to other scientists is very important in scientific inquiry. In this investigation it is important for you to find good ways of showing what you learned to others in your class. Be sure your maps and displays are clearly labeled and well organized.

An icon was developed to draw students' attention to brief descriptions of inquiry processes in the margins of the student module. The icon and explanations provide opportunities to direct students' attention to what they are doing, and thus serve as an important metacognitive tool to stimulate thinking about thinking.

National Science Education Standards link...

Content Standard A
As a result of activities in grades 5-8, all students should develop:
- Abilities necessary to do scientific inquiry
- Understandings about scientific inquiry

Abilities Necessary to do Scientific Inquiry
- Identify questions that can be answered through scientific investigations
- Use appropriate tools and techniques to gather, analyze, and interpret data
- Develop descriptions, explanations, predictions, and models using evidence
- Think critically and logically to make the relationships between evidence and explanations
- Recognize and analyze alternative explanations and predictions
- Communicate scientific procedures and explanations
- Use mathematics in all aspects of scientific inquiry

(From National Science Education Standards, pages 145-148)

Benchmarks for Science Literacy link...

The Nature of Science Inquiry: Grades 6 through 8
- At this level, students need to become more systematic and sophisticated in conducting their investigations, some of which may last for several weeks. That means closing in on an understanding of what constitutes a good experiment. The concept of controlling variables is straightforward, but achieving it in practice is difficult. Students can make some headway, however, by participating in enough experimental investigations (not to the exclusion, of course, of other kinds of investigations) and explicitly discussing how explanation relates to experimental design.

- Student investigations ought to constitute a significant part—but only a part—of the total science experience. Systematic learning of science concepts must also have a place in the curriculum, for it is not possible for students to discover all the concepts they need to learn, or to observe all of the phenomena they need to encounter, solely through their own laboratory investigations. And even though the main purpose of student investigations is to help students learn how science works, it is important to back up such experience with selected readings. This level is a good time to introduce stories (true and fictional) of scientists making discoveries – not just world-famous scientists, but scientists of very different backgrounds, ages, cultures, places, and times.

(From Benchmarks for Science Literacy, page 12)

Teacher's Edition

8. Digging Deeper

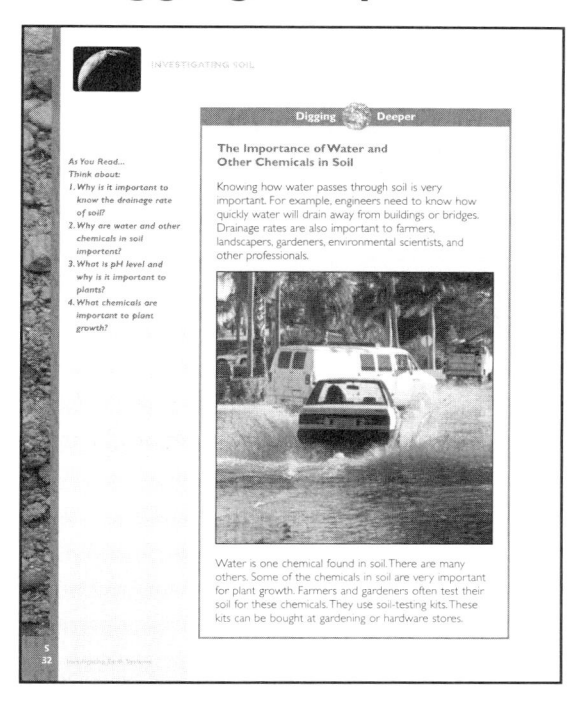

This section provides text, illustrations, data tables, and photographs that give students greater insight into the concepts explored in the activity. Teachers often assign **As You Read** questions as homework to guide students to think about the major ideas in the text. Teachers can also select questions to use as quizzes, rephrasing the questions into multiple choice or "true/false" formats. This provides assessment information about student understanding and serves as a motivational tool to ensure that students complete the reading assignment and comprehend the main ideas.

This is the stage of the activity that is most appropriate for teachers to explain concepts to students in whole-class lectures or discussions. References to **Blackline Masters** are available throughout the Teacher's Edition. They refer to illustrations from the textbook that teachers may photocopy and distribute to students or make overhead transparencies for lectures or presentations.

9. Review and Reflect

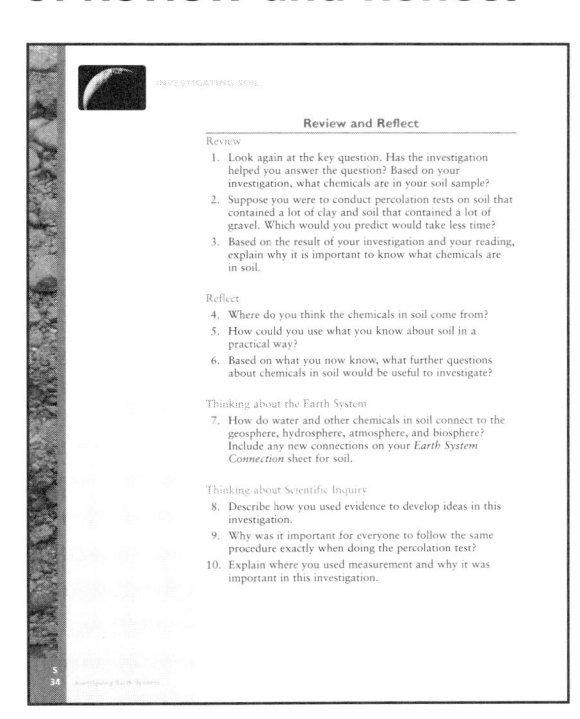

Questions in this feature ask students to use the key principles and concepts introduced in the activity. Students are sometimes presented with new situations in which they are asked to apply what they have learned. The questions in this section typically require higher-order thinking and reasoning skills than the **As You Read** questions. Teachers can assign these questions as homework, or have students complete them in groups during class. Assigning them as homework economizes time available in class, but has the drawback of making it difficult for students to collectively revisit the understanding that they developed as they worked through the concepts as a group

Investigating Earth Systems: Soil   **XXV**

during the investigation. A third alternative is, of course, to assign the work individually in class. When students work through application problems in class, teachers have the opportunity to interact with students at a critical juncture in their learning – when they may be just on the verge of "getting it."

Review and Reflect prompts students to think about what they have learned, how their work connects with the Earth system, and what they know about scientific inquiry. Another one of the important principles of learning used to guide the selection of content in *Investigating Earth Systems* was that:

"To develop competence in an area of inquiry, students must (a) have a deep foundation of factual knowledge, (b) understand facts and ideas in the context of a conceptual framework, and (c) organize knowledge in ways that facilitate retrieval and application." (*How People Learn: Bridging Research and Practice* National Research Council, 1999, P. 12.)

Reflecting on one's learning and one's thinking is an important metacognitive tool that makes students examine what they have learned in the activity and then think critically about the usefulness of the results of their inquiry. It requires students to take stock of their learning and evaluate whether or not they really understand "how it fits into the Big Picture." It is important for teachers to guide students through this process with questions such as "What part of your work demonstrates that you know and can do scientific inquiry? How does what you learned help you to better understand the Earth system? How does your work contribute or relate to the concepts of the Big Picture at the end of the module?"

10. Final Investigation: Putting It All Together

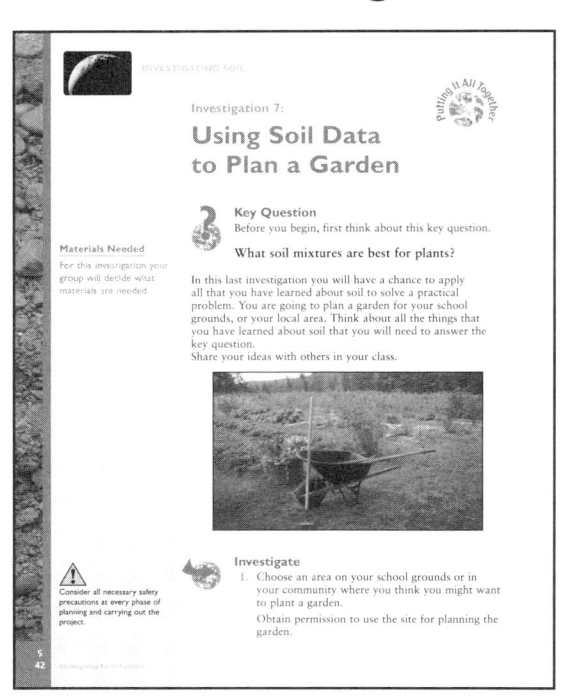

In the final investigation in each *Investigating Earth Systems* module, your students will apply all the knowledge they have about the topics explored to solve a practical problem or situation. Requiring students to apply all they have gained toward a specific outcome should serve as the main assessment information for the module. A sample assessment rubric is provided in the back of this Teacher's Edition. Whatever rubric you employ, it is important that you share this with students at the outset of the final investigation so that they understand the criteria upon which their work will be judged.

Teacher's Edition

The instructions provided to students are purposely open-ended, but can be completed to various levels depending upon how much knowledge students apply. During the final investigation, your role is to be a participant observer, moving from group to group, noticing how students go about the investigation and how they are applying the experience and understanding they have gained from the module.

11. Reflecting

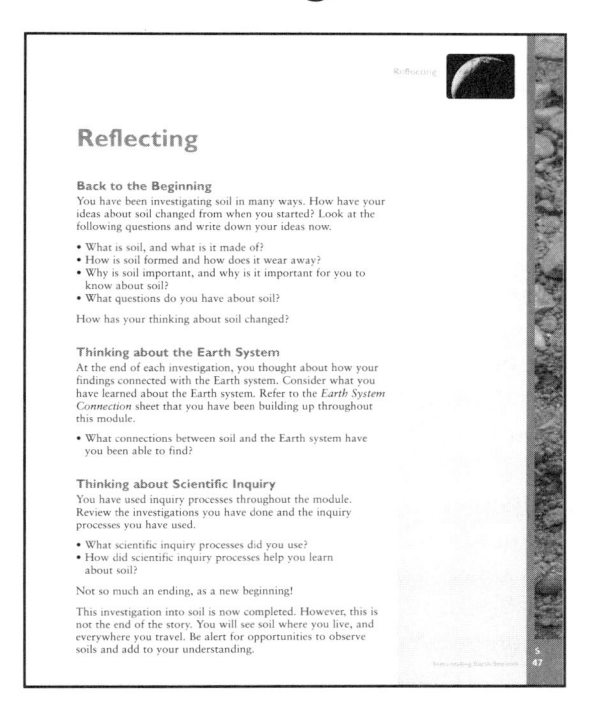

Now that students are at the end of the module, they are provided with questions that ask them to reflect upon all that they have learned about Earth science, inquiry, and the Earth system. The first set of questions (**Back to the Beginning**) are the same questions used in the pre-assessment. Teachers often ask students to revisit their initial responses and provide new answers to demonstrate how much they have learned.

12. The Big Picture

The five key concepts below underlie Earth science in general and *Investigating Earth Systems* in particular. Collectively, the nine modules in the *Investigating Earth Systems* series are designed to help students understand each of these concepts by the time they complete grade 8. Many of the concepts that underlie the Big Picture may be difficult for students to grasp easily. As students develop their ideas through inquiry-based investigations, you can help them to make connections with these key scientific concepts. As a reminder of the importance of the major understandings, the Student Book has a copy of the Big Picture in the back of the book near the **Glossary**.

Be on the lookout for chances to remind students that:
- Earth is a set of closely linked systems.
- Earth's processes are powered by two sources: the Sun and Earth's own inner heat.
- The geology of Earth is dynamic, and has evolved over 4.5 billion years.
- The geological evolution of Earth has left a record of its history that geoscientists interpret.
- We depend upon Earth's resources—both mined and grown.

13. Glossary

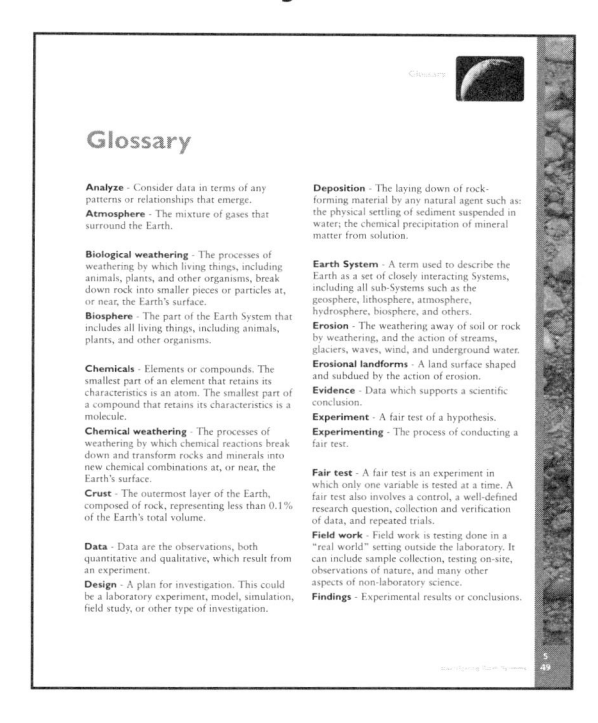

Words that may be new or unfamiliar to students are defined and explained in the **Glossary** of the Student Book. Teachers use their own judgment about selecting the terms that appear in the **Glossary** that are most important for their students to learn. Teachers typically use discretion and consider their state and local guidelines for science content understanding when assigning importance to particular vocabulary, which in most cases is very likely to be a small subset of all the scientific terms introduced in each module and defined in the **Glossary**.

References

How People Learn: Bridging Research and Practice (1999) Suzanne Donovan, John Bransford, and James Pellegrino, editors. National Academy Press, Washington, DC. 78 pages. The report is also available online at www.nap.edu.

Using the Investigating Earth Systems Web Site

www.agiweb.org/ies

The *Investigating Earth Systems* web site has been designed for teachers and students.
- Each *Investigating Earth Systems* module has its own web page that has been designed specifically for the content addressed within that module.
- Module web sites are broken down by investigation and also contain a section with links to relevant resources that are useful for the module.
- Each investigation is divided into materials and supplies needed, **Background Information,** and links to resources that will help you and your students to complete the investigation.

Enhancing Teacher Content Knowledge

Each *Investigating Earth Systems* module has a specific web page that will help teachers to gather further **Background Information** about the major topics covered in each activity. An example from one of the IES module web pages is shown below.

Example from an IES Module web page.

Beginning to Investigate Rocks and Landforms

> **To learn more about different types of rocks, visit the following web sites:**
>
> • What are the basic types of rock?, Rogue Community College
> This site lists the basic descriptions of sedimentary, metamorphic and igneous rocks. Detailed information on each type of rock is also available.
> (http://www.jersey.uoregon.edu/~mstrick/AskGeoMan/geoQuerry13.html)
>
> *1. Sedimentary Rocks:*
> • Sedimentary Rocks, University of Houston
> Detailed description of the composition, classification, and formation of sedimentary rocks.
> (http://ucaswww.mcm.uc.edu/geology/maynard/INTERNETGUIDE/appendf.htm)
> • Image Gallery for Geology, University of North Carolina
> See more examples of sedimentary rocks.
> (http://www.geosci.unc.edu/faculty/glazner/Images/SedRocks/SedRocks.html)
> • Sedimentary Rocks Laboratory, Georgia Perimeter College
> Read a thorough discussion of clastic, chemical, and organic sedimentary rocks. Illustrations accompany each description.
> (http://www.gpc.peachnet.edu/~pgore/geology/historical_lab/sedrockslab.php)
> • Textures and Structures of Sedimentary Rocks, Duke University
> View a collection of slides of different sedimentary rocks as either outcrops or thin sections viewed through a microscope.
> (http://www.geo.duke.edu/geo41/seds.htm)
> • Sedimentary Rocks, Washington State University
> Learn more about sedimentary processes, environments of deposition in relation to different sedimentary rocks. Topics covered include depositional environments, chemical or mechanical weathering, deposition and lithification, and classification.
> (http://www.wsu.edu/~geology/geol101/sedimentary/seds.htm)

Obtaining Resources

The inquiry focus of *Investigating Earth Systems* will require teachers to obtain local or regional maps, rocks, and data. The *Investigating Earth Systems* web site helps teachers to find such materials. The web page for each *Investigating Earth Systems* module provides a list of relevant web sites, maps, videos, books, and magazines. Specific links to sources of these materials are often provided.

Managing Inquiry in Your Investigating Earth Systems Classroom

Materials

The proper management of materials can make the difference between a productive, positive investigation and a frustrating one. If your school has purchased the materials kit (available through It's About Time Publishing) most materials have been supplied. In many cases there will be additional items that you will need to supply as well. This can include photocopies or transparencies (**Blackline Masters** are available in the **Appendix**), or basic classroom supplies like an overhead projector or water source. On occasion, students will bring in materials. If you do not have the materials kit, a master list of materials for the entire module precedes the first investigation. Tips on using and managing materials accompany each investigation.

Safety

Each activity has icons noting safety concerns. In most cases, a well-managed classroom is the best preventive measure for avoiding danger and injury. Take time to explain your expectations before beginning the first investigation. Read through the investigations with your students and note any safety concerns. The activities were designed with safety in mind and have been tested in classrooms. Nevertheless, be alert and observant at all times. Often, the difference between an accident and a calamity is simple monitoring.

Time

This module can be completed in six weeks if you teach science in daily 45-minute class periods. However, there are many opportunities to extend the investigations, and perhaps to shorten others. The nature of the investigations allows for some flexibility.

An inquiry approach to science education requires the careful management of time for students to fully develop their investigative experience and skills. To help you manage time, each activity in the module comes with a matrix that breaks activities down into parts. It is designed to help you think about what you might accomplish with your students in 40 minutes of working class time, and what you might consider assigning for homework.

Most investigations will not easily fit into one 45-minute lesson. You may feel it necessary to extend them over two or more class periods. Some investigations include long-term studies. Where this is the case you may need to allow time for data collection each day, even after moving on to the next investigation.

Classroom Space

On days when students work as groups, arrange your classroom furniture into small group areas. You may want to have two desk arrangements—one for group work and one for direct instruction or quiet work time.

The Student Journal

The student journal is an important component of each *IES* module. (See the **Appendix** in this Teacher's Edition for a **Blackline Master** of the Journal–Entry Checklist.) Your students are young scientists as they investigate Earth science questions. Encourage your students to think of themselves in this way and to see their journals as records of their investigations.

The journal serves other functions as well. It is a key component in performance assessment, both formative and summative. (Formative evaluation involves the ongoing evaluation of students' level of understanding and their development of skills and attitudes. Summative evaluation is designed to determine the extent to which instructional objectives have been achieved for a topic.) Encourage your students to record observations, data, and experimental results in their journals. Answers to **Review and Reflect** questions at the end of each investigation should also be recorded in the journal. It is very important that students have enough time to review, reflect, and update their journals at the end of each investigation.

Frequent feedback is essential if students are to maintain good journals. This is difficult but not impossible. For many teachers, the prospect of collecting and grading anywhere from 20 to over 100 journals in a planning period, then returning them the next day, seems prohibitive. This does not need to be the case. If you use a simple rubric, and collect journals often, it is possible to grade 100 journals in an hour. It may not be necessary to write comments every time you collect journals; in some cases, it is equally effective to address trends in student work in front of the whole class. For example, students will inevitably turn in journals that contain no dates and/or headings. This leaves many questions unanswered and makes their work very hard to interpret. There is no need to write this comment over and over again! You might want to consider keeping your own teacher journal for this module. This makes a great template for evaluating student journals. In addition to documenting class activities, you might want to make notes on classroom management strategies, materials and supplies, and procedural modifications. Sample rubrics are included in the **Appendix**.

Student Collaboration

The National Science Education Standards, and Benchmarks for Science Literacy emphasize the importance of student collaboration. Scientists and others frequently work in teams to investigate questions and solve problems. There are times, however, when it is important to work alone. You may have students who are more comfortable working this way. Traditionally, the competitive nature of school curricula has emphasized individual effort through grading, "honors" classes, and so on. Many parents will have been through this experience themselves as students and will be looking for comparisons between their children's performance and other students. Managing collaborative groups may therefore present some initial problems, especially if you have not organized your class in this way before.

Below are some key points to remember as you develop a group approach.

- Explain to students that they are going to work together. Explain *why* ("two heads are better than one" may be a cliché—but it is still relevant).
- Stress the responsibility each group member has to the others in the group.
- Choose student groups carefully to ensure each group has a balance of ability, special talents, gender and ethnicity.
- Make it clear that groups are not permanent and they may change occasionally.
- Help students see the benefits of learning with and from each other.
- Ensure that there are some opportunities for students to work alone (certain activities, writing for example, are more efficiently done in solitude).

Student Discussion

Encourage all students to participate in class discussions. Typically, several students dominate discussion while others hesitate to volunteer comments. Encourage active participation by explicitly stating that you value all students' comments. Reinforce this by not rejecting answers that appear wrong. Ask students to clarify contentious comments. If you ask for students' opinions, be prepared to accept them uncritically.

Teacher's Edition

Assessing Student Learning in Investigating Earth Systems

The completion of the final investigation serves as the primary source of summative assessment information. Traditional assessment strategies often give too much attention to the memorization of terms or the recall of information. As a result, they often fall short of providing information about students' ability to think and reason critically and apply information that they have learned. In *Investigating Earth Systems*, the solutions students provide to the final investigation in each module provide information used to assess thinking, reasoning, and problem-solving skills that are essential to life-long learning.

Assessment is one of the key areas that teachers need to be familiar with and understand when trying to envision implementing *Investigating Earth Systems*. In any curriculum model, the mode of instruction and the mode of assessment are connected. In the best scheme, instruction and assessment are aligned in both content and process. However, to the extent that one becomes an impediment to reform of the other, they can also be uncoupled. *Investigating Earth Systems* uses multiple assessment formats. Some are consistent with reform movements in science education that *Investigating Earth Systems* is designed to promote. **Project-based assessment**, for example, is built into every *Investigating Earth Systems* culminating investigation. At the same time, the developers acknowledge the need to support teachers whose classroom context does not allow them to depart completely from "traditional" assessment formats, such as paper and pencil tests.

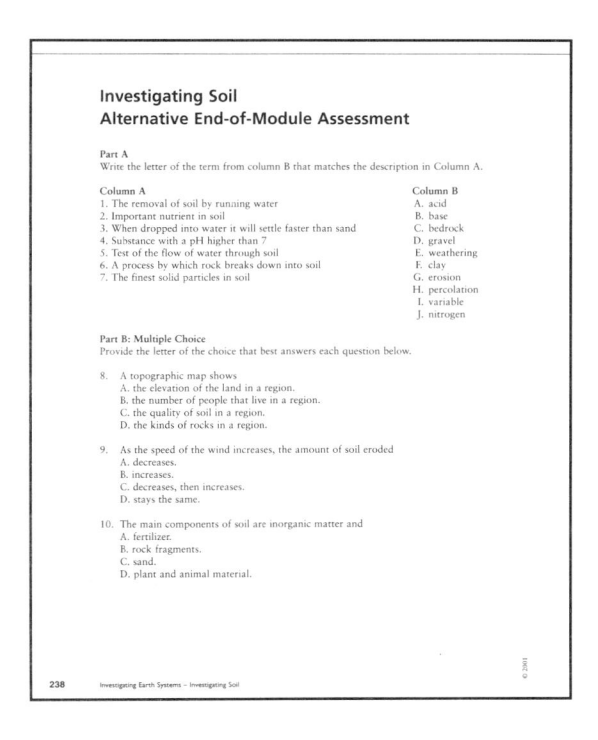

In keeping with the discussion of assessment outlined in the National Science Education Standards (NSES), teachers must be careful while developing the specific expectations for each module. Four issues are of particular importance in that they may present somewhat new considerations for teachers and students. These four issues are dealt with on the next two pages.

Investigating Earth Systems: Soil   XXXV

1. Integrative Thinking

The National Science Education Standards (NSES) state: "Assessments must be consistent with the decisions they are designed to inform." This means that as a prerequisite to establishing expectations, teachers should consider the use of assessment information. In *Investigating Earth Systems*, students often must be able to articulate the connection between Earth science concepts and their own community. This means that they have to integrate traditional Earth science content with knowledge of their surroundings. It is likely that this kind of integration will be new to students, and that they will require some practice at accomplishing it. Assessment in one module can inform how the next module is approached so that the ability to apply Earth science concepts to local situations is enhanced on an ongoing basis.

2. Importance

An explicit focus of NSES is to promote a shift to deeper instruction on a smaller set of core science concepts and principles. Assessment can support or undermine that intent. It can support it by raising the priority of in-depth treatment of concepts, such as students evaluating the relevance of core concepts to their communities. Assessment can undermine a deep treatment of concepts by encouraging students to parrot back large bodies of knowledge-level facts that are not related to any specific context in particular. In short, by focusing on a few concepts and principles, deemed to be of particularly fundamental importance, assessment can help to overcome a bias toward superficial learning. For example, assessment of terminology that emphasizes deeper understanding of science is that which focuses on the use of terminology as a tool for communicating important ideas. Knowledge of terminology is not an end in itself. Teachers must be watchful that the focus remains on terminology in use, rather than on rote recall of definitions. This is an area that some students will find unusual if their prior science instruction has led them to rely largely on memorization skills for success.

3. Flexibility

Students differ in many ways. Assessment that calls on students to give thoughtful responses must allow for those differences. Some students will find the open-ended character of the *Investigating Earth Systems* module reports disquieting. They may ask many questions to try to find out exactly what the finished product should look like. Teachers will have to give a consistent and repeated message to those students, expressed in many different ways, that the ambiguity inherent in the open-ended character of the assessments is an opportunity for students to show what they know in a way that makes sense to them. This also allows for the assessments to be adapted to students with differing abilities and proficiencies.

4. Consistency

While the module reports are intended to be flexible, they are also intended to be consistent with the manner in which instruction happens, and the kinds of inferences that are going to be made about students' learning on the basis of them. The *Investigating Earth Systems* design is such that students have the opportunity to learn new material in a way that places it in context. Consistent with that, the module reports also call for the new material to be expressed in context. Traditional tests are less likely to allow this kind of expression, and are more likely to be inconsistent with the manner of teaching that *Investigating Earth Systems* is designed to promote. Likewise, in that *Investigating Earth Systems* is meant to help students relate Earth Science to their community, teachers will be using the module reports as the basis for inferences regarding the students' abilities to do that. The design of the module reports is intended to facilitate such inferences.

An assessment instrument can imply but not determine its own best use. This means that *Investigating Earth Systems* teachers can inadvertently assess module reports in ways that work against integrative thinking, a focus on important ideas, flexibility in approach, and consistency between assessment and the inferences made from that assessment.

All expectations should be communicated to students. Discussing the grading criteria and creating a general rubric are critical to student success. Better still, teachers can engage students in modifying and/or creating the criteria that will be used to assess their performance. Start by sharing the sample rubric with students and holding a class discussion. Questions that can be used to focus the discussion include: Why are these criteria included? Which activities will help you to meet these expectations? How much is required? What does an "A" presentation or report look like? The criteria should be revisited throughout the completion of the module, but for now students will have a clearer understanding of the challenge and the expectations they should set for themselves.

Investigating Earth Systems Assessment Tools

Investigating Earth Systems provides you with a variety of tools that you can use to assess student progress in concept development and inquiry skills. The series of evaluation sheets and scoring rubrics provided in the back of this Teacher's Edition should be modified to suit your needs. Once you have settled on the performance levels and criteria and modified them to suit your particular needs, make the evaluation sheets available to students, preferably before they begin their first investigation. Consider photocopying a set of the sheets for each student to include in his or her journal. You can also encourage your students to develop their own rubrics. The final investigation is well-suited for such, since students will have gained valuable experience with criteria by the time they get to this point in the module. Distributing and discussing the evaluation sheets will help students to become familiar with and know the criteria and expectations for their work. If students have a complete set of the evaluation sheets, you can refer to the relevant evaluation sheet at the appropriate point within an *IES* lesson.

1. Pre-Assessment

The pre-assessment activity culminates with students putting their journals together and adding their first journal entry. It is important that this not be graded for content. Credit should be given to all students who make a reasonable attempt to complete the activity. The purpose of this pre-assessment is to provide a benchmark for comparison with later work. At the end of the module, the central questions of the pre-assessment are repeated in the section called **Back to the Beginning**.

Teacher's Edition

2. Assessing the Student Journal

As students complete each investigation, reinforce the need for all observations and data to be organized well and added to the journals. Stress the need for clarity, accurate labeling, dating, and inclusion of all pertinent information. It is important that you assess journals regularly. Students will be more likely to take their journals seriously if you respond to their work. This does not have to be particularly time-consuming. Five types of evaluation instruments for assessing journal entries are available at the back of this Teacher's Edition to help you provide prompt and effective feedback. Each one is explained in turn below.

Journal–Entry Evaluation Sheet

This sheet provides you with general guidelines for assessing student journals. Adapt this sheet so that it is appropriate for your classroom. The journal entry evaluation sheet should be given to students early in the module, discussed with students, and used to provide clear and prompt feedback.

Journal–Entry Checklist

This checklist provides you and your students with a guide for quickly checking the quality and completeness of journal entries. You can assign a value to each criterion, or assign a "+" or "-" for each category, which you can translate into points later. However you choose to do this, the point is to make it easy to respond to students' work quickly and efficiently. Lengthy comments may not be necessary. Depending on time constraints, you may not have time to write comments each time you evaluate journals. The important thing is that students get feedback—they will do better work if they see that you are monitoring their progress.

Key–Question Evaluation Sheet

This sheet will help students to learn the basic expectations for the warm-up activity. The **Key Question** is intended to reveal students' conceptions about the phenomena or processes explored in the activity. It is not intended to produce closure, so your assessment of student responses should not be driven by a concern for correctness. Instead, the evaluation sheet emphasizes that you want to see evidence of prior knowledge and that students should communicate their thinking clearly. It is unlikely that you will have time to apply this assessment every time students complete a warm-up activity, yet in order to ensure that students value committing their initial conceptions to paper and taking the warm-up seriously, you should always remind students of the criteria. When time permits, use this evaluation sheet as a spot check on the quality of their work.

Investigation Journal–Entry Evaluation Sheet

This sheet will help students to learn the basic expectations for journal entries that feature the write-up of investigations. *IES* investigations are intended to help students to develop content understanding and inquiry abilities. This evaluation sheet provides a variety of criteria that students can use to ensure that their work meets the highest possible standards and expectations. When assessing student investigations, keep in mind that the **Investigate** section of an *IES* lesson corresponds to the explore phase of the learning cycle (engage, explore, apply, evaluate) in which students explore their conceptions of phenomena through hands-on activity. Using and discussing the evaluation sheet will help your students to internalize the criteria for their performance. You can further encourage students to internalize the criteria by making the criteria part of your "assessment conversations" with them as you circulate around the classroom and discuss student work. For example, while students are working, you can ask them criteria-driven questions such as: "Is your work thorough and complete? Are all of you participating in the activity? Do you each have a role to play in solving the problem?" and so on.

Review and Reflect Journal–Entry Evaluation Sheet

Reviewing and reflecting upon one's work is an important part of scientific inquiry and is also important to learning science. Depending upon whether you have students complete the work individually or within a group, the **Review and Reflect** portion of each investigation can be used to provide information about individual or collective understandings about the concepts and inquiry processes explored in the investigation. Whatever choice you make, this evaluation sheet provides you with a few general criteria for assessing content and thoroughness of student work. Adapt and modify the sheet to meet your needs. Consider involving students in selecting and modifying the criteria for evaluating their end of investigation reflections.

3. Assessing Group Participation

One of the challenges to assessing students who work in collaborative teams is assessing group participation. Students need to know that each group member must pull his or her weight. As a component of a complete assessment system, especially in a collaborative learning environment, it is often helpful to engage students in a self-assessment of their participation in a group. Knowing that their contributions to the group will be evaluated provides an additional motivational tool to keep students constructively engaged. These evaluation forms (Group–Participation Evaluation Sheets I and II) provide students with an opportunity to assess group participation. In no case should the results of this evaluation be used as the sole source of assessment data. Rather, it is better to assign a weight to the results of this evaluation and factor it in with other sources of assessment data. If you have not done this before, you may be surprised to find how honestly students will critique their own work, often more intensely than you might do.

4. Assessing the Final Investigation

Students' work throughout the module culminates with the final investigation. To complete it, students need a working knowledge of previous activities. Because it refers back to the previous steps, the last investigation is a good review and a chance to demonstrate proficiency. For an idea on how to use the last investigation as a performance-based exam, see the section in the **Appendix**.

5. Assessing Inquiry Processes

There is an obvious difficulty in assessing individual student proficiency when the students work within a collaborative group. One way to do this is to have a group present its results followed by a question-and-answer session. You can direct questions to individual students as a way of checking proficiency. Another is to have every student write a report on his or her role in the investigation, after first making it clear what this report should contain. Individual interviews are clearly the best option but may not be feasible given the time constraints of most classes.

6. Traditional Assessment Options

A traditional paper-and pencil-exam is included in the **Appendices**. While performance-based assessments may offer teachers more insight into student skill levels, computer-generated tests are also useful—especially so in states with state-sponsored exams. Additionally, some students are strong in one area and not as strong in another. Using a variety of methods for assessing and grading students' progress offers a more complete picture of the success of the student—and the teacher.

Reviewing and Reflecting upon Your Teaching

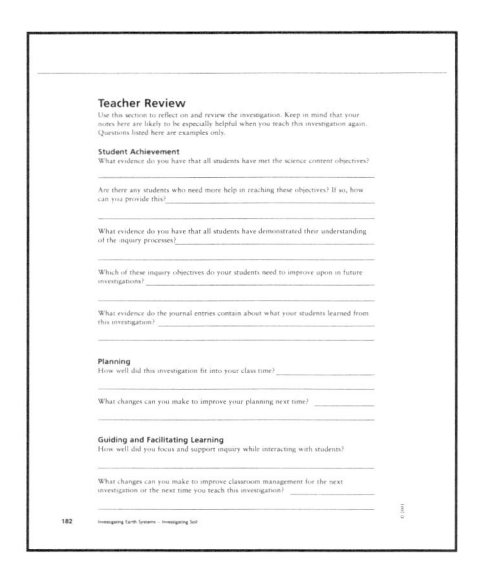

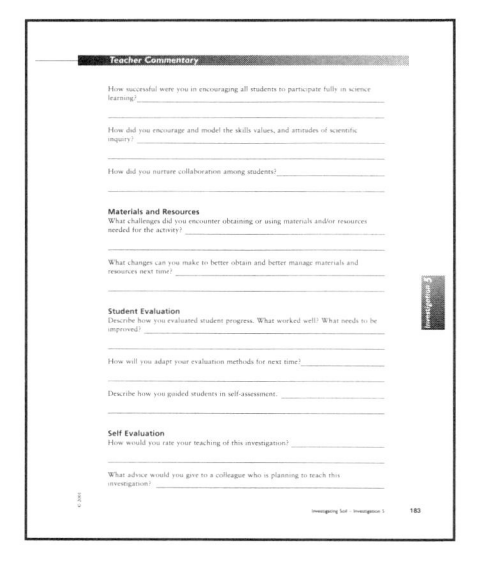

Reviewing and Reflecting upon Your Teaching provides an important opportunity for professional growth. A two-page Teacher Review form is included at the end of each investigation. The purpose of these reviews is to help you to reflect on your teaching of each investigation. We suggest that you try to answer each question at the completion of each investigation, then go back to the relevant section of this Teacher's Edition and write specific comments in the margins. Use the comments the next time you teach the investigation. For example, if you found that you were able to make substitutions to the list of materials needed, write a note about those changes in the margin of that page of this Teacher's Edition.

Investigating Soil: Introduction

The concept of soil varies greatly from discipline to discipline. To the agriculturist, soil is the thin, uppermost layer of the Earth that supports plant growth. To the soil scientist, soil is the layer of material at the Earth's surface that has developed from underlying bedrock by weathering processes. Ordinarily, the agriculturist's soil comprises only the upper part of the soil scientist's soil. To the civil engineer, soil is any material at the Earth's surface that can be excavated without blasting. It usually includes loose material that a soil scientist would not consider soil. In this module we take mostly the soil scientist's view of soil. Soil science, also called pedology (after the Greek *pedon*, meaning ground), is a scientific discipline in itself, but it overlaps a number of other scientific and engineering disciplines: geography, geology, ecology, and environmental science and engineering. Soil scientists study the nature, origin, and classification of soils, and they map the distribution of soils across the continents of the world.

In one sense, soil is a waste product. It is the material that is left over from the decay of bedrock at the Earth's surface. From day to day and from year to year we see little change in the exposed bedrock of the continents, but on much longer time scales the combination of agents such as water, oxygen, acids, and temperature changes break down rocks physically and decompose them chemically to produce soil.

Soils vary enormously in their characteristics from place to place. This can occur for several reasons, most important being original bedrock composition and climate. Also, soils vary in their nature downward from the land surface. Soils pass gradually downward, over distances of a meter or so, into largely unweathered bedrock (or into sediment upon which soil later developed). Over the past century, soil scientists have devoted great effort to classifying soils, mapping their distribution on the continents, and studying the nature of their vertical profiles.

Soil is essential to human civilization, because most of the world's food supply, meats as well as fruits and vegetables, depends upon plant growth in soil (the only exception being seafood). Students may not be aware that the production of beef, chicken, pork, lamb, and other animal meats depends upon plant growth, because the animals feed upon plant materials.

Soils are not permanent. On human time scales, the natural fertility of soil can be reduced by agricultural practices that do not involve replenishment of organic matter and nutrients. Soil erosion is one of the serious environmental problems faced by human society. Soils develop slowly from fresh rock and mineral material over time scales of hundreds of years, at the shortest, to many thousands or even tens of thousands of years. Soils must therefore be considered to be a nonrenewable resource. In many parts of the world, including the U.S., soils are being eroded by wind and running water much faster than they are being formed.

Soils can become buried, and sometimes preserved through geologic time, by later sedimentation. This is common, for example, on a river floodplain. Sedimentary geologists who study ancient sedimentary rocks in the geological record often use ancient soil horizons (called paleosols) in a succession of sedimentary rocks to help interpret the physical and chemical environment in which the soil originally formed.

Soils (which, in the language of Earth systems, could be called the pedosphere) are an important component of Earth systems. They are the active interface between the atmosphere and hydrosphere, on the one hand, and the geosphere, on the other hand. On land, the biosphere owes its rich diversity to soil. Plants root in the soil, macroorganisms feed upon the plant material (or on other macroorganisms that themselves feed upon the plant material), and microorganisms make their homes in the soil in astounding numbers. Carbon dioxide in the atmosphere is taken up by plants and fixed in their tissue as carbon-bearing organic compounds. When the organic material decays, it releases carbon dioxide, which passes upward through the soil back into the atmosphere. Gases can be dissolved in water (like carbon dioxide gas in soft drinks). Some of the carbon dioxide gas released into the atmosphere gets dissolved in liquid water droplets in the atmosphere. When carbon dioxide gas combines with water, a weak acid is formed:

$$CO_2 + H_2O \rightarrow H_2CO_3(aq)$$
$$\text{carbon dioxide} \quad \text{water} \quad \text{carbonic acid}$$

This acid then breaks down into very reactive hydrogen ions and bicarbonate ions:

$$H_2CO_3 \rightarrow H^+ + HCO_3^-$$
$$\text{carbonic acid} \quad \text{hydrogen ion} \quad \text{bicarbonate ion}$$

This reaction further speeds up the rate of soil weathering. Rainwater and atmospheric gases are actively transported, both upward and downward, through the soil layer. Oxygen, in water and as atmospheric oxygen, contributes to the further chemical breakdown of soil minerals.

Introduction

The mechanisms of soil formation involve many factors, each intricately connected with Earth systems.

Additional Resources

Sustaining Our Soils & Society (1998). American Geological Institute. Thomas E. Loynachan, Kirk W. Brown, Terence H. Cooper, and Murray H. Milford. ISBN 0-922152-50-0. Paperback, 8.50"x11.00" 64 pp. The booklet, *Sustaining Our Soils and Society*, and the colorful poster, *Soils Sustain Life*, offer new tools for raising awareness and understanding of our soil resources.

The *Investigating Earth Systems* web site www.agiweb.org/ies also contains a variety of links to web sites that will help you deepen your understanding of content and prepare you to teach this module on *Investigating Soil*.

Students' Conceptions about Soil

In one sense, almost everybody knows something about what soil is. If you live in a rural area, students may be more likely to know about soil and soil properties. In suburban areas, students will have had experience with gardens where plants are put into soil, watered and weeded. Even in urban environments, people can still see the soil where plants carve out a living, sometimes in surprising places. In addition, many people keep plants in potted soil inside or around their homes. Consequently, soil is something about which most students will have ideas. However, their understanding of soil may be limited to everyday experience. Some may simply associate the word with "dirt."

Middle school students have a very limited understanding of the way that soil forms through natural processes of chemical weathering, physical weathering, and the action and/or decomposition of animals and plants that live on and in soil. Many students associate humans with the formation of soil. This may be due, in part, to having seen "potting soil" in stores, which helps students to associate soil with people that produce soil for sale. Some students conceive of soil as dirt that people or animals have added things to (fertilizer, manure, compost, and so on).

Field testing of the program revealed that middle school students have a variety of ideas about soil. Not all students have the same ideas, but the following sampling of responses will give you a general sense of what you might expect middle school students to know about soil coming into the module. A more extensive sampling of responses is presented in the pre-assessment section of this Teacher's Edition.

Ideas about Properties of Soil
- Soil is dirt with a lot of things in it.
- Soil is solid.
- Soil is a mixture of fragments of rock, sand, and dirt.

Ideas about What is in Soil
Soil has a lot of nutrients in it.
Soil has rocks and shells in it.
Soil has chemicals in it.

Ideas about How Soil Forms
Soil forms near water where sediment is formed and carried away.
Soil forms when dirt, bark from a tree, and minerals are combined.
Soil is made when people add things to the ground.

Ideas about Why Soil is Important
Soil can support plant life or is used to grow things.
There are different kinds of soil with different uses.
Soil keeps plants alive, and without trees there is not enough air from trees.

Questions Students Have about Soil
Who really made soil? How do companies manufacture soil?
How does soil get its minerals?
Can people speed up the production of soil?

It is crucial that you find out what informal ideas your students already have about soil before beginning this module. The pre-assessment activity will tell you much of what you need to know in addressing your students' unique needs.

Introduction

Investigating Soil: Module Flow

Activity Summaries	Emphasis
Pre-Assessment Students describe their understanding of key concepts explored in the module.	Recording initial content knowledge and understandings.
Introducing Soil Students discuss their ideas and experiences related to the topics they will be investigating.	Putting the investigations into a meaningful context.
Investigation 1: Beginning to Investigate Soil Students observe a soil sample and formulate questions about soil to investigate.	Forming questions for inquiry. Modeling investigative approach.
Investigation 2: Separating Soil by Settling Students separate soil samples using air and water.	Making observations, predicting experimental outcomes.
Investigation 3: Separating Soil by Sieving Students further separate soil by using strainers of different sizes.	Using alternative methods to conduct an experiment.
Investigation 4: Examining Core Samples of Soil Students analyze a core sample of soil and compare it to other samples.	Using tools to gather and interpret data and applying mathematics in an investigation.
Investigation 5: Water and Other Chemicals in Soil Students do soil percolation tests and use test kits to measure nutrients in soil.	Using measurement and tools in an investigation. Assessing data and recording findings.
Investigation 6: Soil Erosion Students model how wind and water erode soil.	Using inquiry processes to design and test models.
Investigation 7: Using Soil Data to Plan a Garden Students use all their knowledge about soil to plan a garden suited for their school's soil profile.	Synthesizing results, applying results of previous experiments, and communicating findings.
Reflecting Students review the science content and inquiry processes they used throughout the module.	Assessing student learning.

Investigating Soil: Module Objectives

Investigation	Science Content	Inquiry Process Skills
Investigation 1: Beginning to Investigate Soil Students observe a soil sample and formulate questions about soil to investigate.	Students will collect evidence that: 1. There are many items in soil. 2. Items in soil can be physically separated. 3. Soil is part of the Earth's Systems.	Students will: 1. Make observations using the senses. 2. Record observations in a variety of ways. 3. Use tools (such as magnifiers) to make observations. 4. Formulate a question to investigate. 5. Devise a plan to investigate a question. 6. Communicate observations and findings to others.
Investigation 2: Separating Soil by Settling Students separate soil samples using air and water.	Students will collect evidence that: 1. Items in soil can be physically separated by settling through air and water. 2. Objects on or near the Earth are pulled toward Earth's center by the gravitational force. 3. Friction is a force that acts on particles as they fall through air and water.	Students will: 1. Make observations using the senses. 2. Record observations in a variety of ways. 3. Formulate a question to investigate. 4. Predict possible outcomes of the investigation. 5. Devise a plan (fair test) to investigate a question. 6. Communicate observations and findings to others.
Investigation 3: Separating Soil by Sieving Students further separate soil by using strainers of different sizes.	Students will collect evidence that: 1. Most soils contain many kinds of material. 2. Items in soil can be physically separated by a variety of methods. 3. Soil is a mixture of organic and inorganic matter. 4. Inorganic material in soil includes rock and mineral particles of various sizes.	Students will: 1. Make observations using the senses. 2. Collect and review data using tools. 3. Use evidence to develop ideas. 4. Communicate observations and findings to others.

Introduction

Investigating Soil: Module Objectives

Investigation	Science Content	Inquiry Process Skills
Investigation 4: Examining Core Samples of Soil Students analyze a core sample of soil and compare it to other samples.	Students will collect evidence that: 1. Soil is often layered. 2. Soil layers have particular characteristics (color, texture, etc.). 3. Soil can contain both living and non-living components. 4. Soil composition varies depending on location. 5. Soil is part of the Earth's System.	Students will: 1. Follow a protocol to collect core samples of soil. 2. Use map skills to record where soil samples were collected. 3. Use tools (such as magnifiers) to make observations about soil core samples. 4. Use their senses to make observations about soil core samples. 5. Make a record of the nature of the soil core sample using words, tables, and pictures. 6. Organize findings into a display. 7. Find evidence to support conclusions. 8. Communicate data and conclusions to others. 9. Compare conclusions between groups. 10. Generate new questions to investigate about soil samples and their sources.
Investigation 5: Water and Other Chemicals in Soil Students do soil percolation tests and use test kits to measure nutrients in soil.	Students will collect evidence that: 1. The nature of soil allows water to percolate through it. 2. Particle size affects the rate at which water percolates through soil. 3. Soil can be analyzed for nutrient content and pH levels. 4. Drainage rates and soil chemistry affect plants that live in soil.	Students will: 1. Collect a sample according to a set procedure. 2. Predict an event. 3. Design and conduct an investigation. 4. Organize and present findings to others. 5. Use findings from a laboratory test to explain a field study (percolation). 6. Use evidence to verify or refute a prediction. 7. Follow a set procedure for a laboratory test. 8. Compare test results to standards. 9. Interpret test results in terms of their practical implications.

Investigating Soil: Module Objectives

Investigation	Science Content	Inquiry Process Skills
Investigation 6: Soil Erosion Students model how wind and water erode soil.	Students will model how: 1. Wind and water that move across a soil surface apply forces to soil particles. 2. The stronger the water current or wind, the more particles that are put into motion. 3. Proper planning can reduce soil erosion. 4. Soil erosion is a serious problem in many areas. 5. Predictions can be derived from models and these predictions can be tested.	Students will: 1. Identify questions that can be answered through scientific investigations. 2. Design and conduct a scientific investigation. 3. Use appropriate tools and techniques to gather, analyze and interpret data. 4. Develop descriptions, explanations, predictions, and models using evidence.
Investigation 7: Using Soil Data to Plan a Garden Students use all of their knowledge about soil to plan a garden suited to their school's soil profile.	Students will collect evidence that: 1. Plants depend on the water content in soil to survive. 2. Different plants require different amounts of soil drainage to thrive. 3. Plants are affected by the nutrients in soil. 4. Soil type varies with location. 5. Soil nutrients can be altered through biological and chemical means. 6. Soil is part of the Earth's Systems.	Students will: 1. Use sampling and testing skills to analyze soil from a local area. 2. Compare soil data to resource information about plant needs. 3. Devise methods to alter growing conditions for plants, if necessary. 4. Devise a garden plan based upon soil test results, needs of different plants, costs, and available methods of altering growing conditions. 5. Create a scale diagram of the garden plan. 6. Communicate their plan to others. 7. Arrive at a consensus on the most effective garden plan. 8. Create a presentation of the garden plan for the school administration. 9. Present the garden plan. 10. Carry out the garden plan. 11. Analyze the garden plan.

Introduction

National Science Education Content Standards

Investigating Earth Systems is a Standards-driven curriculum. That is, the scope and sequence of the series is derived from, and driven by, the National Science Education Standards (NSES) and the American Association for the Advancement of Science (AAAS) Benchmarks for Science Literacy (BSL). Both specify content standards that students should know by the completion of eighth grade.

Unifying Concepts and Processes
- Systems, order, and organization
- Evidence, models, and explanation
- Constancy, change, and measurement
- Evolution and equilibrium

Science as Inquiry
- Identify questions that can be answered through scientific investigations
- Design and conduct a scientific investigation
- Use tools and techniques to gather, analyze, and interpret data
- Develop descriptions, explanations, predictions, and models based upon evidence
- Think critically and logically to make the relationships between evidence and explanation
- Recognize and analyze alternative explanations and predictions
- Communicate scientific procedures and explanations
- Use mathematics in all aspects of scientific inquiry
- Understand scientific inquiry

Physical Science
- Properties and changes of properties in matter
- Transfer of energy

Earth and Space Science
- Structure of the Earth system
- Earth's history

Science and Technology
- Abilities of technological design
- Understandings about science and technology

Science in Personal and Social Perspectives
- Natural hazards
- Science and technology in society

History and Nature of Science
- Science as a human endeavor
- Nature of science

Key NSES Earth Science Standards Addressed in *IES* Soil

A major goal of science in the middle grades is for students to develop an understanding of Earth (and the solar system) as a set of closely coupled systems. The idea of systems provides a framework in which students can investigate the four major interacting components of the Earth System–geosphere (crust, mantle, and core), hydrosphere (water), atmosphere (air), and the biosphere (the realm of living things).

1. Soil consists of weathered rocks and decomposed organic material from dead plants, animals, and bacteria. Soils are often found in layers, with each having a different chemical composition and texture.
2. Water is a solvent. As it passes through the water cycle it dissolves minerals and gases, and carries them to the oceans.
3. Living organisms have played many roles in the Earth System, including affecting the composition of the atmosphere, producing some types of rocks, and contributing to the weathering of rocks.
4. Some changes in the solid Earth can be described as the "rock cycle." Old rocks at the Earth's surface weather, forming sediments that are buried, then compacted, heated, and often recrystallized into new rock. Eventually, those new rocks may be brought to the surface by the forces that drive plate motions, and the rock cycle continues.
5. The Earth processes we see today, including erosion, movement of the lithospheric plates, and changes in atmospheric composition, are similar to those that occurred in the past.

Key AAAS Earth Science Benchmarks Addressed in *IES* Soil

1. The benefits of the Earth's resources–such as fresh water, air, soil, and trees–can be reduced by using them wastefully or by deliberately or inadvertently destroying them. The atmosphere and the oceans have a limited capacity to absorb wastes and recycle materials naturally. Cleaning up polluted air, water, or soil, or restoring depleted soil, forests, or fishing grounds can be very difficult and costly.
2. The Earth is mostly rock. Three-fourths of its surface is covered by a relatively thin layer of water (some of it frozen), and the entire planet is surrounded by a relatively thin blanket of air.
3. Although weathered rock is the basic component of soil, the composition and texture of soil and its fertility and resistance to erosion are greatly influenced by plant roots and debris, bacteria, fungi, worms, insects, rodents, and other organisms.

Introduction

Materials and Equipment List for Investigating Soil

Pre-Assessment
Each group of students will need:
- poster board, poster paper, or butcher paper
- student journal cover sheet (one for each student)

Teachers will need:
- overhead projector, blackboard, or flipchart paper
- transparency of **Blackline Master** Soil P.1 (Questions about Soil) available at the back of this Teacher's Edition

Materials Needed for Each Group per Investigation

Investigation 1
- soil sample (taken from your local area)
- one or more hand lenses
- tools for separating soil (plastic knife, tweezers, tongue depressor, drinking straw)
- white paper
- cup of water
- paper towels

Investigation 2
- 2 clear-plastic 300 mL (10 oz.) cups
- soil sample (taken from your local area)
- 250 mL (8 oz.) cup of water
- stopwatch, or watch with second hand
- plastic drinking straw or stir stick
- disposable latex gloves

Investigation 3
- soil sample (taken from your local area)
- large piece of white poster board
- 4 squares of white poster board (about 10 cm square)
- 3 large mixing bowls
- plastic strainer, with about 2-mm diameter holes
- kitchen sieve, with about 0.5-mm diameter holes
- large plastic cups
- plastic spoon
- hand lens

Investigating Soil – Introduction 11

Investigation 4
- map of sampling area
- piece of 2.5 cm (1") heavy duty PVC pipe about 25 cm (10") long
- metric ruler
- wooden block
- hammer
- garden or work glove
- piece of wooden dowel 30 cm (12") long that fits inside PVC pipe
- plastic wrap
- masking tape and marker (or other method of labeling core sample)
- colored pencils
- plastic knife
- hand lens
- tweezers or tongue depressor

Investigation 5
- garden shovel
- watering can
- supply of water
- 4 clear-plastic (10 oz.) cups with small hole in bottom
- 4 clear-plastic (6 oz.) cups
- cup of sand
- cup of gravel
- cup of clay
- large soup can with ends removed (extras are recommended)
- wooden block
- hammer
- garden or work glove
- timing device
- soil-testing kit

Investigation 6
- stream table or large plastic tray with end cut open
- soil sample to fill the tray
- tube or hose connected to a water faucet
- large garbage pail
- small piece of wood or other item to prop up tray
- large sponge
- metric ruler
- large sample of dry soil, about 5 kg (10 lb.)

Introduction

- electric window fan, or a fan on a stand
- plastic drop cloth
- weights to keep the drop cloth in place
- dust masks

Investigation 7

For this investigation each group will decide what materials are needed.

General Supplies

Although the investigations can be done with the specific materials listed, it is always a good idea to build up a supply of general materials. These include:

- 2 or 3 large clear plastic storage bins about 30 cm x 45 cm x 30 cm deep, with lids (these can be used both for storage and also make good water containers)
- 2 or 3 plastic buckets and one large water container (camping type with a faucet)
- rolls of masking tape, duct tape, and clear adhesive tape
- rolls of plastic wrap and aluminum foil
- clear self-locking plastic bags (various sizes)
- ball of string and spools of sewing thread
- pieces of wire (can be from wire coat hangers)
- stapler, staples, paper clips, and binder fasteners
- safety scissors and one sharp knife
- cotton balls, tongue depressors
- plastic and paper cups of various sizes
- empty coffee and soup cans, empty boxes and egg cartons
- several clear plastic soda bottles (various sizes)
- poster board, overhead transparencies, tracing paper, and graph paper
- balances and/or scales, weights, spring scales
- graduated cylinders, hot plates, microscopes
- safety goggles
- disposable latex gloves
- lab aprons or old shirts
- first aid kit

Pre-assessment

Overview

During the pre-assessment phase, the students complete an open-ended survey of their knowledge and understanding of key concepts explored in the Soil module. Students are given four questions to consider and their responses become the first entry in their journal.

Preparation and Materials Needed

This pre-assessment activity does not appear in the Student Book. Yet it is crucial that you conduct this pre-assessment before introducing this module to your students and distributing the student books. When you complete the pre-assessment activity, you will have important data that tell you what your students already know about soil.

The pre-assessment should not be presented as a test. Make sure that your students are clear about this. Tell students that, at the end of their investigations, they will be able to compare how their ideas and knowledge about soil has changed as a result of their investigations, and that you will also be able to gauge how successful the investigations have been for everyone.

After the pre-assessment and before distributing the student books, take some time to reflect on the ideas your students have. This is the starting point. You need to ensure that what follows fits with your students' prior knowledge.

Materials:
- poster board, poster paper, or butcher paper
- overhead projector, blackboard, or flipchart paper
- overhead transparency of questions (**Blackline Master** *Soil* P.1)
- student journal cover sheet, one for each student (**Blackline Master** *Soil* P.2)

Suggested Teaching Procedure

1. Let students know that what they write in this exercise will become their first entry in a scientific journal that they will keep throughout the module. Explain that each person is going to write down all the ideas that they have about soil. The reason for this is to provide them and you with a starting point for their investigations into soil. Tell students that when they have finished the module, they will answer these same questions again. This will allow them and you to compare how their knowledge about soil has changed as a result of their investigations.

2. Display the pre-assessment questions on an overhead projector (**Blackline Master** *Soil* P.1). Have students write responses to these questions in their journals. Allow a reasonable amount of time for all students to respond. Circulate around the classroom, prompting students to provide as much detail as possible.

Introduction

- What is soil, and what is it made of?
- How is soil formed and how does it wear away?
- Why is soil important, and why is it important for you to know about soil?
- What questions do you have about soil?

Sample Student Responses

Ideas about Properties of Soil
Soil is dirt with a lot of things in it.
Soil is brown in color, light in color, etc.
Soil can be different colors, different shades of brown, etc.
Soil is made out of minerals.
Soil absorbs water.
Soil can be moist or dry.
Soil can be mud.
Soil can be cracked (dried out, desiccated).
Soil is solid.
Soil can be rough (texture).
Soil feels soft.
Soil has different textures.
Soil is a mixture of fragments of rock, sand, and dirt.

Ideas about What is in Soil
Soil has a lot of nutrients in it.
Bugs (insects) and plant matter are in soil.
Soil has living things in it.
Soil has decomposers in it.
Soil has rocks and shells in it.
Soil can have compost in it.
Animal waste.
Soil has chemicals in it.

Ideas about How Soil Forms
Soil forms near water where sediment is formed and carried away.
Soil forms when dirt, bark from a tree, and minerals are combined.
Soil is made when people add things to the ground.
Soil is made by things that decompose in nature.
Soil is made from manure.

Wind picks up fragments of things and blows them all together, including sand, dirt, rock, tree bark, and dust.
Dust becomes dirt, so soil comes from dirt and particles on the Earth.
Soil is made from the dying bodies of animals, plants, and animal waste.
Soil is made of animals that died and turned into soil.
Soil comes from rocks that break apart on mountains.

Ideas about Why Soil is Important
Soil can support plant life or is used to grow things.
Soil is what you use to plant seeds and grow plants.
There are different kinds of soil with different uses.
Soil keeps plants alive and without trees, there is not enough air from trees.

Questions Students Have about Soil
What is soil supposed to smell like?
Who really made soil?
How do companies manufacture soil?
Why is it called soil?
What is the difference between soil and dirt?
What else can you use soil for besides growing things?
How does fertilizer help soil?
Why is it useful in skin care?
Where does soil come from?
How does soil get its minerals?
Where does soil get its color?
What would we do without soil?
Is soil different from fertilizer?
What is the importance of soil?
Can people speed up the production of soil?
How long has soil existed?
What is soil related to?
What is in soil that promotes plant growth?
Why is soil brown?
Is mulch different than soil?
Who first discovered soil?
What type of people use soil?
Do animals eat soil?
Is there another substance that is like soil?

Introduction

3. Give each student a copy of the student journal sheet (**Blackline Master** *Soil* P.2). Direct students to insert the student journal sheet and their pre-assessment into their journal. Explain that they now have one of the most important tools for this investigation into soil: their own scientific journal.

Teaching Tip

What form will journals take? Using loose-leaf notebook paper in a thin three-ring binder enables students to add or remove pages easily. On the downside, loose-leaf pages are more easily lost and students must maintain a regular supply of paper. If you prefer to have students keep journals in composition notebooks or laboratory notebooks, have them trim the journal cover sheet to the appropriate size and paste it onto the first page of their notebooks.

4. Divide students into groups. Instruct the groups to discuss the following:
   - Ideas we have about soil
   - Questions we have about soil

   One member of the group should record his/her group's ideas and questions on a sheet of poster board, poster paper, or butcher paper.

5. Discuss student responses by having each group, in turn, report on its ideas. As groups are responding, build up two important lists (ideas and questions) for everyone to see (on a chalk board, flipchart paper, poster board, or an overhead transparency).

6. Direct students to add these "Ideas" and "Questions" to their journals.

7. This completes the pre-assessment phase. Distribute copies of *Investigating Soil*.

Assessment Opportunity

It will be useful for you to review what your students have written before moving further into the module. This will alert you, in advance, to any specific problems they may encounter when beginning the module. Keep these lists. They also represent informal pre-assessment data, and you will be able to revisit them with your students at the end of this soil investigation to help track changes in understanding.

Investigating Soil

INVESTIGATING SOIL

The Earth System

The Earth System is a set of systems that work together in making the world we know. Four of these important systems are:

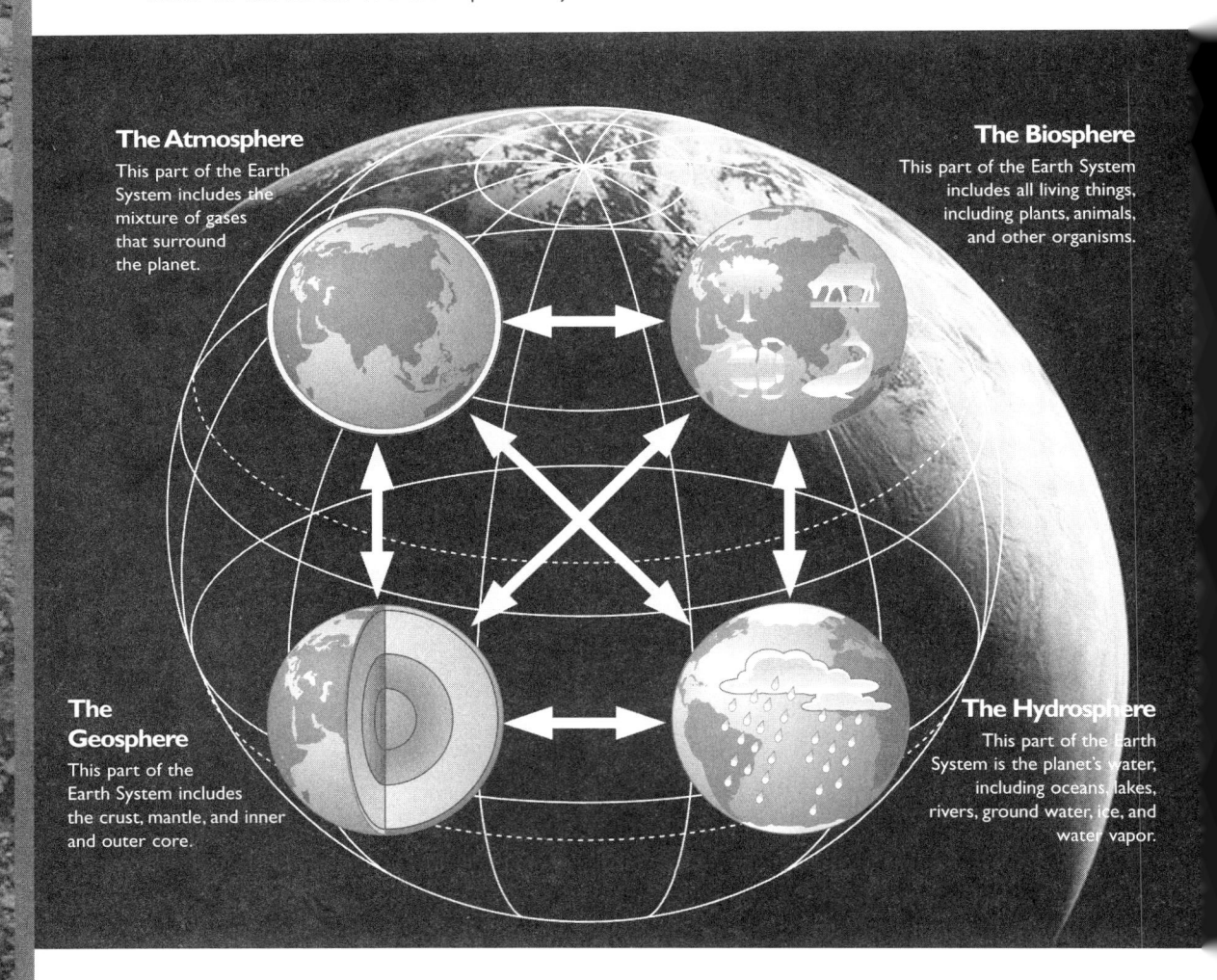

The Atmosphere
This part of the Earth System includes the mixture of gases that surround the planet.

The Biosphere
This part of the Earth System includes all living things, including plants, animals, and other organisms.

The Geosphere
This part of the Earth System includes the crust, mantle, and inner and outer core.

The Hydrosphere
This part of the Earth System is the planet's water, including oceans, lakes, rivers, ground water, ice, and water vapor.

Investigating Earth Systems

Investigating Earth Systems – Investigating Soil

Introduction

Introducing the Earth System

Understanding the Earth system is an overall goal of the *Investigating Earth Systems* series. The fact that Earth functions as a whole and all its parts operate together in meaningful ways to make the planet work as a single unit underlies each module. In each module the students are guided along the way to consider this fundamental principle.

At the end of every investigation, students are asked to link what they have discovered with ideas about the Earth system. Questions are provided to guide their thinking, and they are asked to write their responses in their journals. They are also reminded on occasion to record the information on an *Earth System Connection* sheet. This sheet will provide a cumulative record of the connections that the students find as they work through the investigations in the module. Not all the connections will be immediately apparent to your students. They will probably need your help to understand how some of the things they have been investigating connect to the Earth system. However, as they work through the modules, by the time they complete grade 8, they should have a working knowledge of how they and their environment function as a system within a system, within a system…of the Earth system.

Distribute a copy of the *Earth System Connection* sheet (**Blackline Master** *Soil* I.1) available at the back of this Teacher's Edition. Have the students place the sheet in their journals.

Explain to the students that at the end of each investigation they will be asked to reflect on how the questions and outcomes of their investigation relate to the Earth system. Tell them they should enter any new connections that they discover on their "connection" sheet. Encourage them to also include connections that they have made on their own, that is, do not limit their entries to just those suggested in the **Thinking about the Earth System** questions in **Review and Reflect**. Use the **Review and Reflect** time to direct students' attention to how local issues relate to the questions they have been investigating. By the end of the module students should have as complete an *Earth System Connection* to *Soil* as possible.

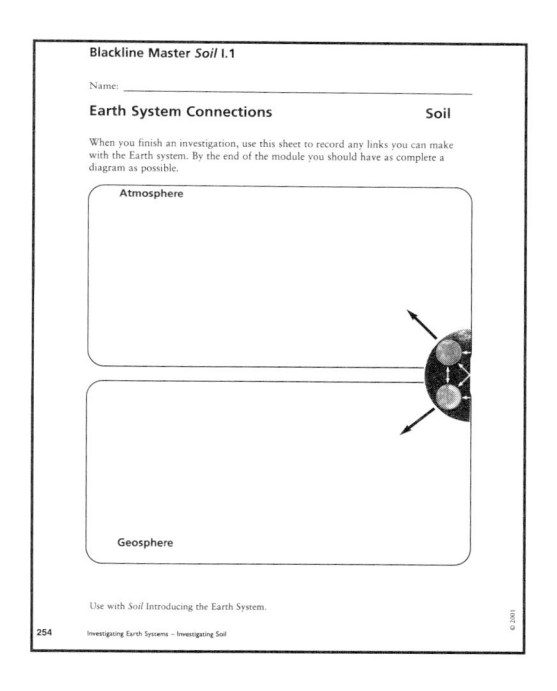

Investigating Soil – Introduction   **19**

Investigating Soil

INVESTIGATING SOIL

Introducing Inquiry Processes

When geologists and other scientists investigate the world, they use a set of inquiry processes. Using these processes is very important. They ensure that the research is valid and reliable. In your investigations, you will use these same processes. In this way, you will become a scientist, doing what scientists do. Understanding inquiry processes will help you to investigate questions and solve problems in an orderly way. You will also use inquiry processes in high school, in college, and in your work.

During this module, you will learn when, and how, to use these inquiry processes. Use the chart below as a reference about the inquiry processes.

Inquiry Processes:	How scientists use these processes
Explore questions to answer by inquiry	Scientists usually form a question to investigate after first looking at what is known about a scientific idea. Sometimes they predict the most likely answer to a question. They base this prediction on what they already know to be true.
Design an investigation	To make sure that the way they test ideas is fair, scientists think very carefully about the design of their investigations. They do this to make sure that the results will be valid and reliable.
Conduct an investigation	After scientists have designed an investigation, they conduct their tests. They observe what happens and record the results. Often, they repeat a test several times to ensure reliable results.
Collect and review data using tools	Scientists collect information (data) from their tests. The data may be numerical (numbers), or verbal (words). To collect and manage data, scientists use tools such as computers, calculators, tables, charts, and graphs.
Use evidence to develop ideas	Evidence is very important for scientists. Just as in a court case, it is proven evidence that counts. Scientists look at the evidence other scientists have collected, as well as the evidence they have collected themselves.
Consider evidence for explanations	Finding strong evidence does not always provide the complete answer to a scientific question. Scientists look for likely explanations by studying patterns and relationships within the evidence.
Seek alternative explanations	Sometimes, the evidence available is not clear or can be interpreted in other ways. If this is so, scientists look for different ways of explaining the evidence. This may lead to a new idea or question to investigate.
Show evidence & reasons to others	Scientists communicate their findings to other scientists to see if they agree. Other scientists may then try to repeat the investigation to validate the results.
Use mathematics for science inquiry	Scientists use mathematics in their investigations. Accurate measurement, with suitable units is very important for both collecting and analyzing data. Data often consist of numbers and calculations.

Investigating Earth Systems

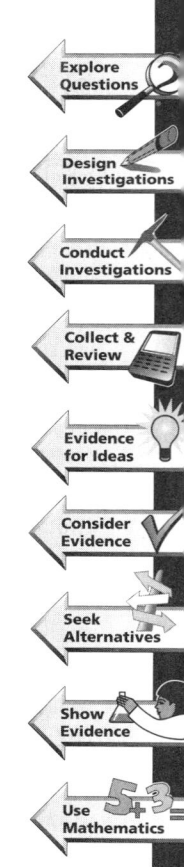

Introduction

Introducing Inquiry Processes

Inquiry is at the heart of *Investigating Earth Systems*. That is why each module title begins with "Investigating." In the National Science Education Standards, inquiry is the first Content Standard. NSES then lists a range of points about inquiry. These fundamental components of inquiry were written into the list shown at the beginning of the student module.

Inquiry depends on active student participation. It is very important to remind students of the steps in the inquiry process as they perform them. Icons that correspond to the nine major components of inquiry appear in the margins of this Teacher Edition. They point out opportunities to teach and assess inquiry understandings and abilities. Stress the importance of inquiry processes as they occur in your investigations. Provoke students to think about why these processes are important. Collecting good data, using evidence, considering alternative explanations, showing evidence to others, and using mathematics are all essential to *IES*. Use examples to demonstrate these processes whenever possible.

At the end of every investigation, students are asked to reflect upon their thinking about scientific inquiry. Refer students to the list of inquiry processes as they answer these questions.

> **Teaching Tip**
>
> If the reading level of the descriptions of inquiry processes is too advanced for some students, you could provide them with illustrations or examples of each of the processes. You may wish to provide students with a copy of the inquiry processes to include in their journals. (**Blackline Master** *Soil* I.2).

INVESTIGATING SOIL

Introducing Soil

Did you ever wonder how soil is formed?

Have you ever seen a dust storm?

Why is there soil in the water?

Have you ever looked at things that live in the soil?

Introduction

Introducing Soil

This is an introduction to the module for your students. It is designed to set their investigations into a meaningful context.

Students will have had a variety of experiences relating to soil. This is an opportunity for them to offer some of their own experiences in a general discussion, using these questions as prompts. Some students may be able to cite additional experiences to those asked for here. Encourage a wide-ranging discussion based on what students are able to offer from their experiences.

Since your students have just spent time in the pre-assessment thinking about and discussing what they already know about soil, it is probably not necessary to have them complete another journal entry. They will be anxious to get to work on their investigations.

You may want to quickly summarize the main points that emerge from the discussion. You could do this on a chalkboard, easel pad, or an overhead transparency. For your own assessment purposes, it will be useful to keep a record of these early indicators of student understanding.

Teaching Tip

The photograph in the upper left shows some of the major components of soil: weathered rock, sand, silt, clay and organic matter. The photograph on the upper right reveals what can happen when the protective cover of vegetation is missing or removed from soil, leaving it exposed to wind and water that can easily remove it. Here, wind strips the finer particles and carries them away. The photograph in the lower left shows how water erodes and transports soil. The removal of soil by the river caused a tree to topple into the stream and gave the water its brown color. The photograph on the lower right shows a plant and dark soil rich in organic matter.

Investigating Soil

INVESTIGATING SOIL

Why Is Soil Important?

Soil may be something that you have only thought of as being under your feet. Soil is much more than that. It is one of Earth's greatest natural resources. Almost all of the world's food crops grow in soil, and the world's livestock eat plants that grow in soil.

Most plants put their roots down into the soil, and many kinds of animals make their homes in soil. Most building foundations are put into the soil. The layer of soil that covers the land is very thin compared to the radius of the Earth. The depth of soil can vary widely on the Earth, from a few centimeters to tens of meters. This is not very much considering the importance of the soil layer to life on Earth!

What Will You Investigate?

To help you understand the science of soil, you will study local soil samples. You will also do research to see what types of garden plants grow best in the soil. Here are some of the characteristics of soil that you will investigate:

- the kinds of materials in soil;
- the arrangement of soil materials;
- the amount of water the soil can hold;
- how water flows through soil;
- how soil is eroded.

You will need to practice your problem-solving skills and be good observers and recorders. You will also need to be creative in finding out information about your local soil. It will be necessary for you to consult a variety of sources. To complete your investigations you will work together with other members of your class.

Investigating Earth Systems

Introduction

Why is Soil Important?

Read (or have a student read) this section aloud. This introduction gives information from which students can conduct their own investigations throughout the module.

You may want to start by having your students read this section carefully, then discuss it in their groups. During their discussions, they can write down a series of questions about soil, or about the investigation itself.

You can then have a discussion with your students about their questions.

Teaching Tip

The gopher in the photograph is one of many organisms whose burrowing activity acts to aerate soil.

What Will You Investigate?

It is important for students to get a sense of where they are headed in the module. Students need your help making sense of the series of investigations in the module and connecting what seem like unrelated investigations into a cohesive network of ideas. Reviewing this section of the introduction is the first step toward constructing a conceptual framework of "the Big Picture" as it is explored in *Investigating Soil*, including change over time and the dynamic nature of the Earth's geology.

This would be a good time to review with students the titles of the activities in the Table of Contents. Ask students to explain how the titles of the activities relate to the descriptions in "What Will You Investigate?"

Discussing the final investigation will help students to understand the overall goal of the module. In the final investigation, students use all they have learned to plan a garden for their school grounds or their local area. Consider introducing students to evaluation rubrics so that they can see how their work will be assessed. Sample rubrics are included in the back of this Teacher's Edition.

NOTES

Teacher Commentary

INVESTIGATION 1: BEGINNING TO INVESTIGATE SOIL

Background Information

1. Soil Formation

Soil forms from bedrock. Loose sediment, itself a product of bedrock, can be transported and deposited by water, wind, or glacier ice. In glaciated areas, including much of the northern U.S., most of the land surface is covered by a layer of glacial sediment. This sediment was most likely deposited either directly from the moving ice or from meltwater streams that flowed out from the terminus of a glacier. In nonglaciated areas, sediment is restricted mainly to major river valleys.

In areas covered with sediment, soils form in the uppermost part of the sediment layer. If you dig a hole or trench in such areas there is a downward transition from weathered sediment (the soil scientist's soil) to unweathered sediment (the civil engineer's soil). Your shovel will strike solid bedrock, usually very abruptly, at some even deeper level in the sediment. In areas without sediment, however, the soil scientist's soil has developed directly upon bedrock, and there is a gradual downward transition from the uppermost soil layer into fresh, solid bedrock. As you dig down through the soil, the material becomes more and more difficult to dig, until blasting becomes necessary. The depth to fresh bedrock varies from less than a meter, in cold, dry climates, to tens or even hundreds of meters, in very warm and humid climates. In these less severe climates, soil-forming processes have had longer times to operate.

All of the loose material at the Earth's surface, whether soil or sediment, has been produced by weathering. Weathering is the term used for the great variety of processes that act at, and very near, the Earth's surface to convert bedrock to loose material. Weathering proceeds at greatly different rates, depending mainly upon climate but also on the inherent "weatherability" of the bedrock. Weathering is too slow to detect from year to year, but changes in bedrock due to weathering are easy to detect within the span of human lifetimes. Substantial changes caused by weathering, however, usually take hundreds to many thousands of years. One notable exception is the rapid weathering of fresh volcanic ash in warm, humid climates, where soils can form from several years to a few decades.

Weathering is traditionally divided into mechanical or physical weathering (the physical breakdown of bedrock into particles) and chemical weathering (the chemical decomposition of the original minerals of the bedrock into new minerals, along with dissolved material). The distinction is somewhat artificial, however, because each kind of weathering tends to facilitate the other.

Physical weathering involves the fracturing of solid bedrock into loose pieces, small and large. Several processes cause fracturing of surface bedrock. In cold regions, the most important is ice wedging, also called frost shattering. Water expands upon freezing. When liquid water in a preexisting crack in a rock freezes, it exerts enormous forces of expansion, causing the fracture to enlarge. In areas with trees, tree roots in cracks in rock slowly enlarge the crack as they grow.

Chemical weathering involves the conversion of certain minerals that formed at great depths and are chemically unstable at the Earth's surface, to other minerals that are stable at the Earth's surface. These changes are particular kinds of chemical reactions. As with many other chemical reactions, they are facilitated by liquid water as the medium for the reactions. Chemical weathering is far slower in dry environments than in humid environments.

Some common minerals, like calcite or gypsum, simply dissolve. The chemical weathering of silicate minerals, however, which are the main constituents of most rocks, is more complicated. Other silicate minerals or oxide minerals are formed, along with dissolved ions like silica, calcium, magnesium, sodium, or potassium. The solid mineral products of chemical weathering are always much finer in size than the original minerals.

Physical weathering and chemical weathering often go hand in hand. An important example is called granular disintegration. It is especially important in rocks like granite that consist of minerals that weather at different rates. Incipient weathering of some of the mineral grains makes them expand slightly, and this expansion causes forces at their boundaries with the more resistant grains. The grains break apart, and fall or roll from the bedrock surface, one by one.

2. Soil Types

What determines the type of soil that forms at a given place on the Earth's surface? The two most important factors, by far, are bedrock type and climate. Your natural assumption would probably be that different rock types produce different soil types. That is true, because different minerals weather to different products. In young soils, those in which soil-forming processes have not had extremely long times to act, soil type is very closely a function of bedrock type. Climate, however, is an important factor as well, in part because the nature of weathering depends on temperature, but even more importantly because it depends on the presence and direction of water movement. In very humid climates, with abundant rainfall, water movement is mostly downward, toward the groundwater table, whereas in dry areas, rainwater first moves downward but is then drawn upward again by capillary action. As a result, the chemical environment in the soils is very different. In old soils, which have become fully developed, climate is usually more important than bedrock composition in determining the soil type, but there are many exceptions. For example, bedrock that consists entirely of quartz can produce nothing but a barren quartz soil, whatever the climate.

Soils vary greatly in their composition, particle size, and vertical structure. For a long time, soil scientists have recognized a great need for systematic classification of soil types. Several schemes for classifying soils have been developed. Some are simple classifications with only a few named categories. One such method is to define soil by the average size of the rock and mineral particles. Sand ranges from 1/16 – 2 mm in diameter. Silt particles are finer than sand and clay particles are smaller than silt. This kind of classification is easy to use, but it necessarily lumps different kinds of soils into the same category. Other classifications are much more detailed. A group of soil scientists from the United States Department of Agriculture created an extremely detailed classification of soils that is widely used by soil scientists around the world. This classification has hundreds and hundreds of named soil types!

Of the hundreds of soil types, two are probably the most common, at least in the United States. Podzol soils, which are common in the eastern U.S., tend to be acidic because there is a net downward

Teacher Commentary

flow of water from the soil surface to the groundwater table. Chernozem soils, which are common in the central and western United States, tend to be alkaline because there is net upward flow of water through the soil profile. The terms *podzol* and *chernozem* were coined by Russian soil scientists over a hundred years ago. Russian soil scientists were pioneers in the scientific study and classification of soils.

The *Investigating Earth Systems* web site www.agiweb.org/ies also contains a variety of links to web sites that will help you deepen your understanding of content and prepare you to teach this investigation.

Investigation Overview

Students begin by considering what they know about soil and writing a question that they could investigate about soil. Their questions are pooled into a class list of possible questions for inquiry. They examine a soil sample and record observations about soil using their sense of sight, touch, and smell. Students select a question to investigate, design an investigation, and, with the approval of their teacher, conduct their investigation into soil. Students present their findings to the class. Text explains how soil forms and the factors that determine the type of soil that forms in a given location.

Goals and Objectives

As a result of this investigation, students will develop a hands-on understanding of the materials in soil and improve their ability to develop questions and design investigations.

Science Content Objectives

Students will collect evidence that:
1. There are many items in soil.
2. Items in soil can be physically separated.
3. Soil is part of the Earth's Systems.

Inquiry Process Skills

Students will:
1. Make observations using the senses.
2. Record observations in a variety of ways.
3. Use tools (such as magnifiers) to make observations.
4. Formulate a question to investigate.
5. Devise a plan to investigate a question.
6. Communicate observations and findings to others.

Connections to Standards and Benchmarks

In this investigation, students will discover some of the directly observable components of soil by examining a soil sample and designing an investigation to answer a question of their own choosing. These observations will start them on the road to understanding the National Science Education Standards shown below.

NSES Links

- An important aspect of this ability [to do scientific inquiry] consists of students' ability to clarify questions and inquiries and direct them toward objects and phenomena that can be described, explained or predicted by scientific investigations.

- Soil consists of weathered rocks and decomposed organic material from dead plants, animals, and bacteria. Soils are often found in layers, with each having a different chemical composition and texture.

Teacher Commentary

AAAS Links

- At this level, students need to become more systematic and sophisticated in conducting their investigations, some of which may last for weeks or more. That means closing in on an understanding of what constitutes a good experiment. The concept of controlling variables is straightforward but achieving it in practice is difficult. Students can make some headway, however, by participating in enough experimental investigations (not to the exclusion, of course, of other kinds of investigations) and explicitly discussing how explanation relates to experimental design.

- Although weathered rock is the basic component of soil, the composition and texture of soil and its fertility and resistance to erosion are greatly influenced by plant roots and debris, bacteria, fungi, worms, insects, rodents, and other organisms.

Preparation and Materials Needed

Preparation

For this activity, you will need a variety of soil samples. If possible, have your students collect these ahead of time, perhaps bringing samples from where they live. This will give them a personal stake in the investigation and will ensure a variety of different soil samples. Alternatively, you could have students collect samples from around the school. If neither of these alternatives is possible, you will need to collect samples yourself.

Quart-size, zip-closing, plastic bags are useful for collecting the soil samples. To best acquire your soil samples, collect topsoil below the leaves in an ungrazed woodland; just below the surface in a pasture or fence row; or in someone's backyard where the soil is damp and relatively undisturbed. Dig to a depth of 6-12 in. to ensure that the soil contains plant and animal material as well as inorganic material.

After putting the soil samples into bags, be sure to label where they were found. Keep them in a cool, moist place. The organisms that are part of the soil can be preserved if the soil is kept in conditions similar to those in which it was found.

Familiarize yourself with the safety precautions for the investigation. Students should have access to disposable latex gloves, lab aprons, and safety goggles while handling soil. Make sure that soap and water are available for students to wash their hands thoroughly after handing soil.

If your students are not used to working in small collaborative groups, you might need to spend some time helping them to understand how to do this. Keep in mind that some students may find this difficult (some prefer to work alone). Help them see that "collaborating" means "working together," with each group member sharing and contributing. Explain that scientists often work in teams. Think about how you will help students to work collaboratively. You may wish to assign roles to each student in their groups. Materials collector, cleanup person, question-asker, and spokesperson are examples of some of the roles students may fill. Emphasize that these roles are not static. A student is not "stuck" in one role for all investigations.

Making Connections ...with the local area

Instruct students to visit a web site that deals with soil in your local area. The IES web site at www.agiweb.org/ies/ will help them with their research. Students can use their research to inform their inquiry in Step 6 of the investigation and incorporate their web research into their presentations in Step 7 of the investigation. Students may wish to give a short presentation or write a short explanation of the site. Collect the site addresses and descriptions. Photocopy the web site list. This will be useful in later investigations.

Teacher Commentary

Materials
- soil sample (taken from your local area)
- one or more hand lenses
- tools for separating soil (plastic knife, tweezers, tongue depressor, drinking straw)
- white paper
- cup of water
- paper towels

Investigating Soil

Investigation 1: Beginning to Investigate Soil

Investigation 1:
Beginning to Investigate Soil

Key Question
Before you begin, first think about this key question.

What can you investigate about soil?

Think about what you already know about soil. Write a question you could investigate about soil based on what you know. Think about what you could do with a soil sample to find an answer to your question.

Share your thinking with others in your group and with your class. Make a list that combines everyone's questions and ideas about things to do. Keep the list for later in this investigation.

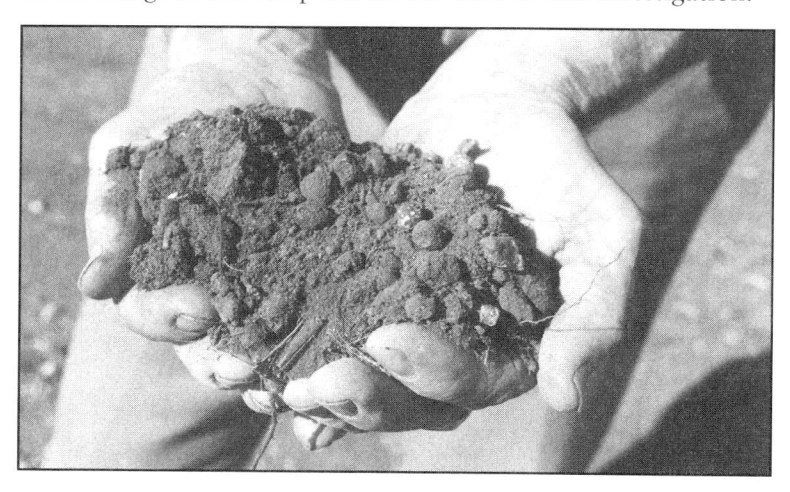

Materials Needed

For this investigation your group will need:

- soil sample (from your local area)
- white paper
- hand lens
- tools for separating soil (plastic knife, tweezers, or tongue depressor)
- cup of water
- paper towels

Investigate
1. Cover a large, flat workspace with white paper.
   Place a soil sample on the paper.
   Divide up the soil so that each member of your group has about a handful.

⚠️ Keep the soil on the paper covering the work area.

Investigating Earth Systems

S1

Teacher Commentary

Key Question

The **Key Question** is a brief warm-up activity designed to elicit students' ideas about the topic explored in the investigation. Emphasize thinking and sharing of ideas. Avoid seeking closure (i.e., the "right answer"). Closure will come through inquiry, reading the text (**Digging Deeper**), by discussing the ideas (lecture), and reflecting on what was learned at the end of the investigation. Make students feel comfortable sharing their ideas by avoiding commentary on the correctness of responses.

Have students read and record their answer to the **Key Question** in their journals. Write the **Key Question** on the board or overhead transparency. Tell students to write the **Key Question** in their journals and to think about and answer the questions individually. Tell them to write as much as they know and to provide as much detail as possible in their responses. Emphasize that the date and the prompt (question, heading, etc.) should be included in journal entries.

Discuss students' ideas. Ask for a volunteer to record responses on the board or overhead projector. This allows you to circulate among the students, encouraging them to copy the notes in an organized way. Urge students to record all the questions in their journals.

Student Conceptions

In this activity, your students will be using their senses to observe a soil sample. Unless they are in a farming community, many of your students may not have looked closely at soil. It is unlikely that they will have separated it into categories. Students may respond that they can investigate the following kinds of questions:
- composition of soil (What's in soil?);
- the mass or density of soil (What does it weigh?);
- regional and local variation of soil (What kinds of soil are there?);
- the effect of soil type on farming, horticulture, and ecology (What's the best type of soil for plants?);
- the importance of soil (How is soil useful?).

Answer for the Teacher Only

This is an opinion question in which students are asked to identify questions that they can investigate about soil. Accept student ideas uncritically at this stage.

Assessment Tool

Key–Question Evaluation Sheet

The **Key–Question Evaluation Sheet** will help students to understand and internalize basic expectations for the warm-up activity. The evaluation sheet emphasizes that you want to see evidence of prior knowledge and that students should communicate their thinking clearly. It is unlikely that you will have time to apply this assessment every time students complete a warm-up activity, yet in order to ensure that students value committing their initial conceptions to paper and taking the warm-up seriously, you should always remind students of the criteria and when time permits, use this evaluation sheet as a spot check on the quality of their work.

Investigate

Teaching Suggestions and Sample Answers

1. Before you distribute materials, review safety procedures. Point out the hand-washing icon shown beside Step 6 on page S3 of the Student Book. Soil contains microorganisms and possibly chemicals that can be a health hazard. Warn students not to taste soil. Ensure they wash their hands with soap and water after handling soil and caution them not to eat or put their hands into their mouths or near their eyes until they have washed.

   Tell students that they will be given a sample of soil to investigate using their senses. Help students to understand that good observations are critical in science and that is very easy to miss an important detail if this is not done fully and systematically.

Teacher Commentary

NOTES

Investigating Soil

INVESTIGATING SOIL

2. Use your senses of seeing, feeling, and smelling (not tasting!) to examine the soil sample. To identify the odor of the soil sample wave your hand over the sample toward your nose.

   a) Record your observations. It is important to record your observations in a way that others can see and understand. One example is shown below. You can adapt this example to suit your needs, or you can choose another method that you think would be better.

⚠ Never place your nose over the sample and inhale. Never taste anything in a science lab. Take care not to rub soil into your eyes.

3. Use the tools to separate and examine the soil more closely.

   a) Add any new observations to your data table or diagram.

Seeing data — Touching data
Other data — Smelling data

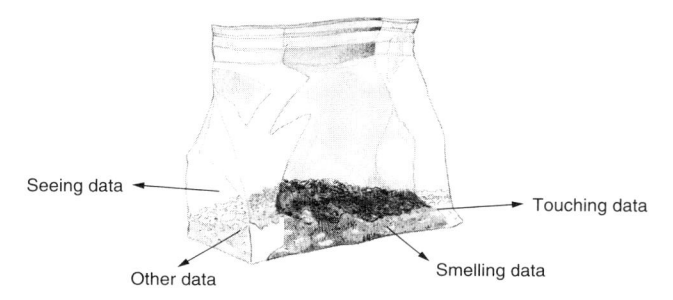

4. Use your data to answer the following questions:

   a) Do you think all groups will have the same findings? Why or why not?

   b) Which questions about soil from your original list have your observations answered? Which questions are still unanswered?

Conduct Investigation

Collect & Review

Evidence for Ideas

Consider Evidence

S2
Investigating Earth Systems

Teacher Commentary

Assessment Tools

Journal–Entry Evaluation Sheet
This sheet provides you with general guidelines for assessing student journals. Adapt this sheet so that it is appropriate for your classroom. The **Journal–Entry Evaluation Sheet** should be given to students early in the module, discussed with students, and used to provide clear and prompt feedback.

Journal–Entry Checklist
This checklist provides you and your students with a guide for quickly checking the quality and completeness of journal entries.

2. Before your students begin investigating their soil, ensure that they have an agreed-upon method of recording their observations. A good way to record observations for this part of the investigation is to make a chart with columns for touching data, seeing data, smelling data, and other data. If you have computers in your classroom, student groups may opt to use one of the word processing options to record their data. Alternatively, students may choose to use or adapt the diagram on this page provided as **Blackline Master** *Soil* 1.1 (Soil Data).

    Observe student inquiry. Circulate among student groups. As you circulate, think of how you can help students develop their inquiry skills. Students are used to observing with the sense of sight. Indeed, most observations are made this way. However, they may not have thought about using their other senses.

    Help students to appreciate the importance of respect for living things. Finding a live insect or other animal may cause some excitement. Animals found could be kept in a clear plastic cup and observed. These living things should be returned to the soil after the investigation.

    a) Observations that your students might make about their soil samples include: variations in texture, color, consistency, presence or absence of animal life (and what types), types and numbers of rocks, types and quantities of plant material, amount of water (in general), and odor.

3. Monitor your students' progress throughout the course of the investigation. Be sure that all students are involved, and that they are recording all data. Some of the materials that students find in their soil samples might surprise them. They may not have appreciated just how many different materials can be in the soil mixture. If different groups are studying soil samples taken from very different places, there might be considerable variation in the materials found.

   a) As you circulate, ask students to share what they have found. Look at students' journals. Provide positive feedback to students who record observations with clarity and detail. In a positive way, point out places where recorded observations might be improved.

> ### Teaching Tip
> You may be surprised at the amount of excitement the observation of soil may generate among your students. Teachers have indicated that some students act as if they have never seen soil before.
>
> Use your judgement and your knowledge of your students as to whether or not you wish to provide them with drinking straws as tools for separating soil. They are valuable tools, however, and they can be misused.

4. Point out that Step 4 asks students to think about the questions from the class discussion. As you circulate, remind students to answer the two questions in Step 4 and to record their responses in their journals.

   a) Student responses will vary. To some degree this depends on what information they have regarding the origin of the soil sample(s). If students know they have different soil samples they will probably believe their samples will be different. This may or may not be true, depending on the degree of variation in local geologic history and topography. More important than their answer is that you elicit ideas for why soil might be different (or similar) from place to place.

   b) It is possible that a few of the students' questions have been answered by observing the soil. If students find answers to their original questions, challenge them to ask new questions. Students will need to have at least one answerable question (using space and materials available). If students experience difficulty with their task, encourage them to be creative.

Teacher Commentary

NOTES

Investigation 1: Beginning to Investigate Soil

5. In your group, decide on one question to investigate further. It might be one of the unanswered questions from your original list, or a new question that has occurred to you from your work so far.

   Prepare a plan to investigate your question. You will need to design an investigation so that it:
   - will lead to an answer (or a part of an answer) to your question;
   - can be done in your working space;
   - can be done safely with the tools and time available;
   - uses a clear method of recording the results.

   a) Write down your plan for investigating your question.

6. With the approval of your teacher, conduct your investigation.

   If necessary, repeat your investigation several times to make sure of your results.

   a) Record your observations and results.

7. Present the findings of your group to the rest of the class.

   Listen to the results of other groups' investigations.

   a) Record and display the findings of all groups.

   b) As a class, what questions do you still think need to be answered?

Inquiry

Scientific Questions

Scientific inquiry starts with a question. Scientists take what they already know about a topic, then form a question to investigate further. The question and its investigation are designed to expand their understanding of the topic. You are doing the same.

Showing Your Findings to Others

One way to present your findings might be for groups to visit each other's working area and observe what each group has done. Another way is for each group to make a presentation in front of the class.

⚠ Have your plan approved by your teacher before you begin your investigation.

Clean up spills immediately.

Teacher Commentary

5. Avoid the temptation to tell students which questions to investigate. Encourage students to write questions that they think can be answered. "Is soil all the same?" is a better question to investigate than "How much soil is there on Earth?" precisely because the former lends itself to classroom inquiry. It is good to encourage questions that require complicated analysis, for this shows ambitious thinking; however, students should also devise at least one question that recognizes the reality of conducting inquiry in their classroom. Emphasize that the date and the prompt (question, heading, etc.) should be included in journal entries.

   Students probably will not have had much experience designing experiments. Remind students that to ensure a valid investigation, students need to consider whether their investigation will answer their question, whether it can be done within the constraints of time and space, whether it can be done safely, and whether they have a clear method of organizing their results. This is an opportunity for students to begin to see that investigations have to be organized within certain limits.

   Inform students that they will be developing a question to investigate, and designing an investigation to find the answer to their question. Tell students that each group will need to agree upon which soil question to investigate and that you must approve this question before they begin. Provide some guidelines for students. Safety issues, the materials you have available, and the amount of time that you can schedule for the investigation are probably the most important factors.

   a) Allow students time to outline their investigation. Circulate among groups and review their plans. If this is the first time that your students have developed a question and/or designed an investigation, you might wish to collect one journal from each group and provide written feedback on the plans.

6. As the groups conduct their investigations, circulate among them, noting how they go about their tasks. Be available for advice or consultation. Look particularly for any instances where a group is testing and refining its procedures, especially if this is improving the validity or reliability of the investigation.

7. When your students have completed their investigations, they will need to communicate their methods and findings to the rest of the class. There are a number of methods for doing this:
   - large poster
   - overhead transparency
   - "visits" by the class to each group in turn as it presents its information
   - oral reports using the chalkboard or chart paper as visual support.

   Instruct each group to nominate a spokesperson. This job should rotate so that all students play this role during the course of the module. Give a set time that each presentation should be so that students stay interested (four minutes works well). The spokesperson should address the following points: a) their group's question, b) their plan, c) their observations, and d) the results of Internet research into local soil (if this was assigned). They should end by discussing whether their group answered the question. Stress that investigations usually lead to more questions–not definitive answers. It is okay if they did not answer their question.

a) In their journals, students should note the questions and outcomes of all groups. Students should record any constructive feedback that they would like to provide to others in their journals. That way, if time does not permit them to share the feedback during the presentation, they can share it with the presenting group at a later date. A scaffold for helping students to think about how to provide constructive feedback is shown below.

b) Students should note and record any questions that remain unanswered. Direct students to ask questions that derive from the investigation they conducted. Student questions should be more focused at this point than the questions they asked at the beginning of the investigation.

Assessment Opportunity
Peer review is an important part of the scientific process. Help students critique their colleagues' presentations by providing sentence stems that they should complete for their reviews. Examples include:
- This presentation was effective at showing…
- This presentation helped me to understand…
- This presentation needed work on…
- To improve this presentation, I would suggest that you…

Assessment Tool
Student Presentation Evaluation Form
The **Student Presentation Evaluation Form** provides simple guidelines for assessing presentations. Adapt and modify the evaluation form to suit your needs. If you decide to use the form, provide it to students and discuss it with them before they begin to prepare their work. Students should always know the expectations for their work before they begin to prepare for an assessment.

Teacher Commentary

NOTES

Investigating Soil

INVESTIGATING SOIL

Digging Deeper

Types of Soil

As You Read...
Think about:
1. What are the main things found in soil?
2. How is soil formed?
3. What determines the type of soil that is formed?

There are many kinds of soil. A group of soil scientists from the U.S. made up a way of grouping soils that is used around the world. This grouping has hundreds of named soil types! All soils, however, are made of just a few main components. Soil consists of fine particles of minerals and rocks, decaying plants, and living plants and animals. You can easily see the larger plants and animals. There are even more tiny plants and animals that you can only see with a microscope.

Soil forms as the solid rock of the Earth, called bedrock, breaks down. It usually takes thousands of years for soil to form from bedrock. In some places, soil forms directly on top of bedrock. In other places, soil forms on a thick layer of loose rock and mineral material. This material, called sediment, has been carried from distant areas by rivers or glaciers.

What determines the type of soil that forms? Only two things are most important: bedrock type, and climate. It should make sense to you that different kinds of bedrock make different kinds of soil. Climate is also important. Water helps chemical reactions in soil to take place. Young soils are soils which have just started to form. In young soils bedrock is more important than climate in determining the type of soil. In old soils, which have become fully formed, climate is usually more important.

Most soils are only a meter or two deep. The nature of the soil changes as you go down. When soil scientists study a soil, they look carefully at the whole thickness of the soil. This type of section of soil is called a soil profile. In a later investigation you will have a chance to study a real soil profile.

Evidence for Ideas

S4
Investigating Earth Systems

Investigating Earth Systems

Teacher Commentary

Digging Deeper

At this stage of the activity, students read a brief passage about soil. This is the ideal time for you to provide students with further information about soil by showing slides, photographs, and through lecture and discussion. Detailed **Background Information** about soil that you can use to develop a lecture is provided at the beginning of each activity in this Teacher's Edition. Reading this information (and visiting the *IES* web site for links to relevant sites) will help you to deepen your own understanding of content and help you prepare to answer questions that students may raise.

The first paragraph on page S4 explains that there are many types of soil and explains the main components of soil. If students have collected soil samples from different locations, they will have already appreciated these facts.

The second paragraph explains two ways that soil forms – on top of solid bedrock and on a layer of loose rock and sediment.

The third paragraph explains that bedrock and climate are the two most important factors that determine the type of soil that forms. Soil that forms on top of granite bedrock will be different than soil that forms on top of basalt bedrock. Climate determines the amount of water that enters the soil and the temperature to which the soil is exposed. Soil generally forms more quickly and is thicker in warm, wet (tropical) climates than in cold, dry (polar) climates.

The final paragraph explains what soil is like from "top to bottom" (the soil profile).

For additional teaching suggestions on how to present the content of this **Digging Deeper** reading section to your students, please refer to the *IES* web site.

About the Photo
Point out the photograph on page S4 and ask students to think about what happens when the river shown in the photograph floods (some of the material that it carries gets washed over the banks of the river and added to the soil).

As You Read…
Instruct students to read the **Digging Deeper** section on page S4, and answer the **As You Read** questions 1–3 (in their journals). Answers to the three questions are shown below.

1. Things found in soil include loose particles of rocks and minerals, decaying plants, living plants and animals and soil moisture and/or air in pore spaces.
2. Soil is formed when bedrock gets broken down by weathering and combined with plant and animal material.
3. The two most important factors in soil formation are bedrock and climate.

Assessment Opportunity

You may wish to select questions from the **As You Read** section to use as quizzes, rephrasing the questions into multiple choice or "true/false" formats. This provides assessment information about student understanding and serves as a motivational tool to ensure that students complete the reading assignment and comprehend the main ideas.

Teacher Commentary

NOTES

Investigating Soil

Investigation 1: Beginning to Investigate Soil

Review and Reflect

Review

Evidence for Ideas

1. What questions about soil were you able to answer from this investigation?
2. What questions about soil do you still want answered?

Reflect

3. What did you expect to find in soil?
4. What surprised you about soil when you looked at it closely?
5. a) What materials did you find in soil?
   b) Where do you think these materials came from?

Thinking about the Earth System

6. Write any connection that you have discovered in this investigation to connect soil to the geosphere, hydrosphere, atmosphere, and biosphere. You can record this information on your *Earth System Connection* sheet.

Thinking about Scientific Inquiry

7. In which part(s) of the investigation did you:
   a) Explore questions to answer by inquiry?
   b) Conduct an investigation?
   c) Design an investigation?
   d) Show evidence and reasons to others?
8. How did you collect and manage your data?
9. What did you do in this investigation to make sure that your results were reliable?

Teacher Commentary

Advise students to review their journals and make sure all entries are complete.

Review and Reflect

Review

The **Review** section is a very important part of the investigation. Here, students need to revisit the **Key Question** "What can you investigate about soil?" and think about how the activity they have engaged in has answered that question. In doing this, students will consolidate and validate their learning. It is important that you allow enough time for this to be done in a thoughtful manner.

1. Answers will vary. Students should have chosen at least one question that they could answer by designing their own investigation.

2. Answers should reflect an understanding of inquiry, namely that one question leads to another.

Reflect

Be sure to give students adequate time to **Review and Reflect** on what they have done and understood in this investigation. Ensure that all students think about and discuss the questions listed here. Be alert to any misunderstandings and, where necessary, help students to clarify their ideas.

3. Students might have expected to find everything from "just dirt" to "insects, rocks, minerals, plant material." Accept all reasonable answers.

4. A good answer might look like this: "There were lots of different-sized soil particles." A less ambitious answer might look like this: "There was more stuff in there than I thought there would be." The difference between a good answer and less ambitious answer is the level of detail.

5. a) Students should incorporate the vocabulary in the **Digging Deeper** reading section. They may have found rocks and minerals, decaying plants, and/or living plants and animals, along with sediment of different-sized particles.

   b) Students should incorporate the vocabulary in the **Digging Deeper** reading section. Good answers will incorporate one or more of the following: bedrock, chemical reactions, or sediment carried by water.

Thinking about the Earth System

It is very important that students begin to relate what they are studying to the wider idea of the Earth System. This is a complex and largely inferred set of concepts that students cannot easily understand from direct observation. Remember, the goal is that students will have a working understanding of the Earth System by the time they complete the eighth grade. Be sure to spend some time helping students make what connections they can between the results of their investigations and the Earth System. A **Blackline Master** (*Earth System Connection* sheet) is available in this Teacher's

Edition. Students can use this to record connections that they make as they complete each investigation.

6. Soil is part of the geosphere, but students may have seen organisms (biosphere component) in the soil. Students should be encouraged to elaborate on their answers. If they saw living organisms, ask them to write what role those organisms play in the soil (decomposing, for example) as well as what role soil plays in the life of the organism (providing shelter, for example). Other connections include: the effect of water (the hydrosphere) in the actions of glaciers and rivers, or the role of wind (the atmosphere) in moving soil from place to place.

Thinking about Scientific Inquiry

The National Science Education Standards include science as inquiry as a Content Standard. This means that students should be learning about scientific processes as a set of content items, in addition to any particular scientific concept (or content) area (in this case the materials in soil). Science as inquiry is a theme that runs through all investigations. Students will need many investigative experiences to grasp the many processes and skills used in scientific investigation. This can be taught as a piece of information, but for a solid understanding, students need considerable firsthand experience in doing it for themselves. Students are asked to reflect on how they used inquiry processes at the end of each investigation. This reflection helps them to build their understanding of "inquiry." Help students understand that they:
- started with a question to investigate;
- had a plan for the investigation;
- collected materials;
- designed a way of recording data;
- carried out an investigation;
- recorded observations and results;
- reached some sensible conclusions;
- shared findings with others.

7. a) Students explored questions at the beginning of the exercise, when discussing "What can you investigate about soil?" Additionally, they formulated a question to answer in class.

    b) Students conducted an investigation by observing soil characteristics and trying to answer their question.

    c) Students designed an investigation to answer a specific research question.

    d) Students presented their findings to the teacher and to other students.

8. Answers will vary. The method should be sufficiently organized and systematic that an outside observer could understand and interpret it.

9. A good answer will include one or more of the following: record data carefully, repeat the steps of the investigation, and compare results with other students.

Teacher Commentary

NOTES

Teacher Review

Use this section to reflect on and review the investigation. Keep in mind that your notes here are likely to be especially helpful when you teach this investigation again. Questions listed here are examples only.

Student Achievement

What evidence do you have that all students have met the science content objectives?
_____

Are there any students who need more help in reaching these objectives? If so, how can you provide this? _____
_____

What evidence do you have that all students have demonstrated their understanding of the inquiry processes? _____
_____

Which of these inquiry objectives do your students need to improve upon in future investigations? _____
_____

What evidence do the journal entries contain about what your students learned from this investigation? _____
_____

Planning

How well did this investigation fit into your class time? _____
_____

What changes can you make to improve your planning next time? _____
_____

Guiding and Facilitating Learning

How well did you focus and support inquiry while interacting with students?
_____

What changes can you make to improve classroom management for the next investigation or the next time you teach this investigation? _____
_____

Investigating Earth Systems – Investigating Soil

Teacher Commentary

How successful were you in encouraging all students to participate fully in science learning? _____
_____

How did you encourage and model the skills values, and attitudes of scientific inquiry? _____
_____

How did you nurture collaboration among students? _____
_____

Materials and Resources

What challenges did you encounter obtaining or using materials and/or resources needed for the activity? _____
_____

What changes can you make to better obtain and better manage materials and resources next time? _____
_____

Student Evaluation

Describe how you evaluated student progress. What worked well? What needs to be improved? _____
_____

How will you adapt your evaluation methods for next time? _____
_____

Describe how you guided students in self-assessment. _____
_____

Self Evaluation

How would you rate your teaching of this investigation? _____
_____

What advice would you give to a colleague who is planning to teach this investigation? _____
_____

NOTES

Teacher Commentary

INVESTIGATION 2: SEPARATING SOIL BY SETTLING

Background Information

1. Gravity and Friction

Separation of particulate materials by settling is a common technique in many fields of science and engineering. In particular, sedimentologists and soil scientists use settling to separate rock and mineral particles by size, especially for fine particles, for which sieving is not practical. The basic idea is simple: large particles fall faster than small particles, and more dense particles fall faster than less dense particles. The reasons for this are not, however, as simple as they might seem at first, and many students are likely to have some confusion about the physical effects that are involved.

By the universal law of gravitation (all matter in the universe attracts all other matter in the universe with a force that varies inversely as the square of the distance between the bodies of matter), the Earth exerts an attracting force on all bodies that are at and above the Earth's surface. This gravity force is expressed as the acceleration that is experienced by a freely falling body, if the friction exerted on the falling body by the surrounding air or water is neglected. (Acceleration is the rate of increase of velocity with time; meters per second, per second for example.) On the Earth's surface, the acceleration of gravity is about 9.8 m/s^2; thus, a falling body increases its fall speed by 9.8 m/s for each second during which it falls (again, if the friction of the surrounding air or water is neglected).

If you could put a feather and a lead ball at the same height in a large chamber from which all the air has been removed by a vacuum pump, and somehow release both the feather and the ball at the same time, their accelerations would be exactly the same, and they would land at the same time! This is inherently counterintuitive to us humans, who live in a sea of air. The reason is not obvious; it has to do with the fact that the mass of the falling body enters into both the universal law of gravitation and also Newton's second law (force equals mass times acceleration) to the same first power.

Friction is the force that arises when one material moves or slides past another material. All materials exert and experience friction, although some materials (like two slippery ice cubes at their melting temperature) exert far less friction on each other than other materials (like two pieces of rough sandpaper). What is important in this investigation is that friction is exerted not only between two solids but also between a solid body and a fluid, like air or water, in which it is immersed.

Everyone knows that moving one's hand through a pool of water takes a certain force. Also, that force is greater when the plane of your hand is perpendicular to the direction of movement than when it is parallel to the direction of movement. That is because there are two different parts to the force that resists the motion: surface friction, and a difference in water pressure between the front of your land and the back of your hand. When your hand is parallel to its movement, the pressure-difference force is small, but when your hand is perpendicular to its movement the pressure-difference force is actually much larger than the friction force. Exactly the same is true of soil particles settling through water.

2. Separating by Settling
Settling in a Fluid

When a body like a soil particle is released in water, it accelerates downward, but as its velocity builds up, the resisting forces (both

Investigating Soil – Investigation 2   57

friction and pressure difference) also build up. Eventually the resisting force builds up to be equal to the weight of the particle, and after that the particle no longer accelerates, but instead falls at a constant velocity. That velocity is called the fall velocity, the settling velocity, or the terminal velocity.

Larger particles have a greater terminal velocity than smaller particles. This is likely to seem obvious to the students, but the reason is not obvious. It is basically because the weight of the particle, which is the force that tends to accelerate the particle downward, varies as the cube of the diameter of the particle, whereas the resisting force, which tends to reduce the acceleration, varies as the square of the diameter of the particle. That is true whether you think about the cross section of the particle, upon which the pressure force depends, or the total surface area of the particle, upon which the friction force depends. The consequence is this: the larger the particle, the greater is its weight relative to its area, and the faster it falls.

A denser body has a larger terminal velocity than a less dense body of the same size. That is because the denser body has a greater weight but the same cross-sectional area and surface area.

Particles with the same size and density fall faster in air than in water. That is because the effective weight of the particle is not its true weight but its submerged weight. Have you ever tried to weigh yourself while standing waist-deep in water? You would weigh less than in air. (And when you are up to your neck in water, the scale would tell you that you weigh almost nothing.) That is because the weight of the water that you displace, from the waist down, is subtracted from your actual weight from the waist down. The same is true of soil particles: the difference in density between the soil particle (with density between 2 g/cm^3 and 3 g/cm^3) and air (with density about 10^{-3} g/cm^3, is much greater than between the soil particle and water (with density about 1 g/cm^3). Important note: the last sentence in the **Digging Deeper** reading section of the Student Edition is erroneous in this regard.

Settling Tubes
Scientists who study sediments and soil materials often use a device called a settling tube to measure the distribution of particle sizes in a sample. The actual terminal velocities of the soil particles are not directly measured. What is done instead is to record the buildup of mass on the bottom of the tube over time, and then, by use of a mathematical transformation, convert that to the distribution of terminal velocities, and finally, knowing independently the relationship between particle size and terminal velocity, convert that to the distribution of particle sizes.

The *Investigating Earth Systems* www.agiweb.org/ies web site also contains a variety of links to web sites that will help you deepen your understanding of content and prepare you to teach this investigation.

Teacher Commentary

Investigation Overview
Students begin by exploring a question about how soil can be separated. They then investigate a method of separating soil into component parts by settling a sample of soil through a column of air. After observing and discussing the results, students are asked to predict what would happen if soil were poured into water. Students test their predictions and collect evidence to evaluate their predictions. The data that students collect provide qualitative data that also help them to better understand the relative sizes and percentages of sizes of material in soil. Students read text that explains how gravity and friction act upon particles that settle through air and water.

Goals and Objectives
As a result of this investigation, students will develop a better understanding of the relative sizes and percentages of components of soil and will improve their ability to make predictions and collect and consider evidence when conducting scientific inquiry.

Science Content Objectives
Students will collect evidence that:
1. Items in soil can be physically separated by settling through air and water.
2. Objects on or near the Earth are pulled toward Earth's center by the gravitational force.
3. Friction is a force that acts on particles as they fall through air and water.

Inquiry Process Skills
Students will:
1. Make observations using the senses.
2. Record observations in a variety of ways.
3. Devise a question to investigate.
4. Predict possible outcomes of the investigation.
5. Devise a plan (fair test) to investigate a question.
6. Communicate observations and findings to others.

Connections to Standards and Benchmarks
In this investigation, students will separate soil into component parts by settling a sample of soil through air and water. These observations will start them on the road to understanding the National Science Education Standards and the American Association for the Advancement of Science (AAAS) Benchmarks shown below.

NSES Links
- Soils are often found in layers, with each having a different chemical composition and texture.
- The motion of an object can be described by its position, direction of motion, and speed.
- An object that is not being subjected to a force will continue to move at a constant speed and in a straight line.

AAAS Link
- Everything on or anywhere near the Earth is pulled toward the Earth's center by gravitational force.

Preparation and Materials Needed

Preparation

You will again need a variety of soil samples. However, this time it may be useful to get these from within the school grounds. In a later investigation, students will be studying the soil layers in this nearby area and, in the final investigation, they will be applying what they know to designing a school-based garden.

Students will be dropping the soil into an empty cup and into a cup of water. You should obtain a soil sample that is fairly crumbly and has a diverse range of particle sizes (fine material to gravel-sized rock fragments). Avoid using a soil sample that is too wet or which consists almost entirely of clay. If the soil is wet, allow it to dry prior to the investigation by leaving the plastic bags open.

Students will be dropping a sample of soil into a cup. The size of the cups is not important; however, the cups must be clear. Newspapers or paper towels should be put down on tables and desks. Keep a broom and mop on hand.

Be sure to have students place their cups in a safe place. If necessary, collect all the cups and place them in a box for overnight storage. Clear sufficient storage space for each group's cup of "soil water." If multiple classes are doing this experiment, make sure to keep each previous class's experiments out of view (and reach!) of subsequent classes.

Familiarize yourself with the safety precautions for the investigation. Students should have access to disposable latex gloves, lab aprons, and safety goggles while handling soil. Make sure that soap and water are available for students to wash their hands thoroughly after handing soil.

Materials
- 2 clear-plastic 300 mL (10 oz.) cups
- soil sample (taken from your local area)
- 250 mL (8 oz.) cup of water
- stopwatch, or watch with second hand
- plastic drinking straw or stir stick

Teacher Commentary

NOTES

Investigating Soil

INVESTIGATING SOIL

Investigation 2:

Separating Soil by Settling

Materials Needed

For this investigation your group will need:

- 2 clear-plastic 300 mL (10 oz.) cups (if your school has tall 1-liter graduated cylinders, they would be even better than plastic cups)
- soil sample from your local area
- 250 mL (8 oz.) cup of water
- stopwatch or watch with second hand
- plastic drinking straw

Key Question

Before you begin, first think about this key question.

How can soil be separated?

In the last investigation you found out that soil is a mixture of many things. Think about how you could separate the materials that make up soil.
Share your ideas with others in your class.

Investigate

1. By making observations, you are going to investigate how soil materials fall through air and through water.

    Place two clear plastic cups side by side on a table.

S6

Investigating Earth Systems

62 Investigating Earth Systems

Teacher Commentary

Key Question

The **Key Question** is a brief warm-up activity designed to elicit students' ideas about the topic explored in the investigation. The **Key Question** is designed to find out what your students know about how soil can be separated.

Write the **Key Question** on the board or overhead transparency. Encourage students to think about what they learned in the first investigation about soil and the kinds of things they found in soil. To encourage participation from all students, tell students to record their ideas in a new journal entry. Emphasize that the date and the prompt (question or context) should be included in journal entries.

Discuss students' ideas. Ask for a volunteer to record responses on the board or overhead projector. This allows you to circulate among the students, encouraging them to copy the notes in an organized way. Encourage students to copy the list in their journals.

Student Conceptions

The most common answer to the **Key Question** will probably be that soil can be separated with a filter. Some students in lower grades may have done an experiment with soil settling, so some students may be familiar with this method.

Answer for the Teacher Only

Soil can be separated a variety of ways. In Investigation 1, students explored how to separate soil into components by pulling the sample apart. In this investigation, students separate soil by settling, dropping the soil through a column of air and then through a column of water. The latter has the added advantage of allowing students to get a sense of the relative amounts of various sizes of material in the soil. The heaviest material (typically the largest material) will settle the fastest, followed by material of decreasing size. Objects less dense than water will float. In the next investigation, students will learn how to separate soil by sieving.

Assessment Tool
Key–Question Evaluation Sheet
The **Key–Question Evaluation Sheet** will help students to understand and internalize basic expectations for the warm-up activity.

About the Photo

You can use the photograph of a river during a flood shown on page S6 to prompt students to think about how soil can be separated. During a heavy rainfall, enormous amounts of soil can be stripped from the land and washed into rivers and streams. The brown color of the water is due to the fine sediment being carried in suspension in the river. This sediment is probably a mixture of sediment normally found along the river and soil that has been eroded and washed into the river. The coarser, heavier sediment is likely to be bouncing along the riverbed. Objects light enough to float are carried at the surface of the river, such as the log seen floating in the photograph. When the river returns to its normal stage, the soil and sediment particles that have been carried by the river will settle out onto the land surface. The coarser particles will settle first and the finer particles will settle out last. Students will observe this same distribution of particle sizes in their investigation.

Teacher Commentary

Investigate

Teaching Suggestions and Sample Answers

Introducing the Investigation

Point out that Steps 2 – 8 require students to record observations and make predictions. These should go in the journal along with a prompt that makes clear what they are observing and/or predicting.

As before, assign roles to each student in their groups (e.g., materials collector, cleanup person, note-taker, and spokesperson). Try to assign different roles than were assigned in Investigation 1.

Remind students about safety procedures. Point out appropriate safety protocols for handling soil, including the use of gloves, safety goggles, and lab aprons. In addition, they should always wash their hands after handling soil (an icon to remind you and students about this is found in the margin of page S8). The drawing on page S7 showing students working sets a good example to discuss with students.

Remind students that good observations are critical in science and that it is very easy to miss an important detail if the observer is not careful. Good scientific observations are complete and made systematically.

This is a student-driven investigation. Once you feel comfortable that students understand the basic safety procedures and any other special instructions, you should allow students to complete the investigation in their groups and circulate around the room to offer assistance and feedback on their work. Observe your students as they investigate. Monitor your students' progress throughout the course of the investigation. Ensure that all students are involved, and that they are recording all data in an organized way. Watch for students who might be having difficulty in participating fully in the collaborative learning process, or those who may be "taking over" the investigation. Help all to see the benefits of working together.

1. Provide soil samples and materials to student groups.

> **Assessment Tool**
> **Journal–Entry Checklist**
> This checklist provides you and your students with a guide for quickly checking the quality and completeness of journal entries.

Investigating Soil

Investigation 2: Separating Soil by Settling

Conduct Investigations

2. If some of the soil is stuck together in lumps, gently break up the lumps with your fingers before starting the investigation.

   Crouch down so that you can see the cups from the side.

Collect & Review

   Have one person very gently pour about 60 mL (1/4 cup) of soil into one of the plastic cups.

   Observe how the soil materials fall through the air inside the cup.

   Repeat the pouring and observing two more times. Take turns pouring and observing.

   a) Record all the observations you can make about how soil falls through air. It may be useful to make a drawing to show your observations.

   b) Describe how the soil looks when it reaches the bottom of the cup.

Explore Questions

3. Discuss this question in your small group: What would happen to the soil if it were poured into water instead of air? Why is that?

   Based on what you already know, make a prediction.

   a) Write down your prediction.

   b) Write down the reason for your prediction.

4. Fill the second plastic cup 3/4 full of water.

Conduct Investigations

   When the water is still, gently pour about 60 mL (1/4 cup) of the soil into the cup and observe.

   Repeat your observations two more times, starting each time with a fresh cup and clear water.

Collect & Review

   a) What happens when the soil meets the water?

   b) What happens to the soil as it passes through the water?

   c) What happens when the soil reaches the bottom of the cup?

Inquiry
Making Predictions

Making a prediction is not just guessing. Predictions are made based on all the information and evidence you already have. When making your prediction, you need to be clear about the reasoning that supports it. That is why giving the reason for your prediction is very important.

Investigating Earth Systems

Teacher Commentary

2. Observe students as they complete Steps 2 through 4. In Step 2, ensure that students spend the time to break up the soil. Show students how they can accomplish this by rolling the soil around with your fingers while it is still in the bag.

   In Step 2 students follow a set procedure. The procedure is an example of what they could do in designing their own fair tests. This is an opportunity for you to emphasize the importance of making careful observations. Repeating this procedure several times should help students increase the accuracy of their observations.

   > **Teaching Tip**
   > Recording Observations
   > Few hints are given on how to record the observations. Use this as an opportunity for students to decide on the best way to record. This might be best done just after they have followed the procedure for the first time, so that they have some idea of what happens. Making drawings with labels might be a good method of recording.

   a) A good observation is that lighter material (plant material, for example) tends to fall slower than heavy material (like rock pieces).

   Sample student observations:
   - When we dropped soil into the cup, everything fell to the bottom of the cup. The clump of soil was on the bottom and leaves and plant materials were on the top.
   - Without water, the soil fell quickly, it fell in clumps, and it isn't in layers. It's all a big mess.

   b) A good response is that light, small objects are on top and heavy large objects are on the bottom. Instruct students to write descriptions that show critical thinking. An answer like "it's all spread out," for example, does not show more than superficial observation.

3. Make sure that students write down their prediction. There is no "right answer" as far as predictions go, but students must be able to back up their prediction with a reason.

4. The exact amount of water poured into the cup is not important. Observe students to ensure that they pour in the soil in a way that minimizes splashing contents. Have them wipe up spills.

Teaching Tip
Helping Students Make Predictions
The idea of making a prediction based on reasoning may be new for many students. Help all to understand this important principle–that predicting is not simply guessing but is based upon careful reasoning. Point out the description in the margin of page S7 (**Inquiry – Making Predictions**). As you circulate among the students, look at their journal entries. Are they writing down their predictions and their reasons for their predictions? Are they labeling their work properly so that, days or weeks from now, the entry will be easily understandable?

a) When soil hits water, students should see that some of the soil stops or slows at the surface. Some of these particles remain at or near the surface, while others fall at different rates. The largest heaviest particles go straight to the bottom.

b) Students should note that the mixture is separated into sub-types.

c) By the time the soil reaches the bottom of the cup, it has become visibly differentiated.

Sample student observations for Step 4:
- When we dropped the soil into water, the material settled according to sizes of rocks and minerals. Fine grains of sand or dirt settled on the top of the soil. All of the leaves, grass, and twigs and bits of acorns floated on the top of the water. The rocks and clumps of soil were on the bottom.
- When soil was dropped into water, it fell more slowly. It separated. All of the sand is on the bottom, in the middle it looks like watery root beer, and the roots and bark are floating on the top.

Teaching Tip
Helping Students to Observe Closely
Soil landing on, and passing through, water is a slower process than soil passing through air. However, what happens in the separating process is much more complicated. Encourage your students to observe everything closely. Your students are observing the behavior of soil as it falls through two fluids, air, and water. They will notice that the gaseous fluid (air) did not separate the soil as well as the water did. Ask students to think about the differences between air and water that could explain why this is so. Water, being a liquid, will offer greater resistance to the soil particles as they descend through it.

Teacher Commentary

NOTES

INVESTIGATING SOIL

Inquiry
The Importance of Evidence

The word **evidence** may be familiar to you in criminal investigations. However, evidence is also important in science. In science, valid conclusions depend on evidence that can be trusted. Others should be able to do the same experiment and come up with the same evidence.

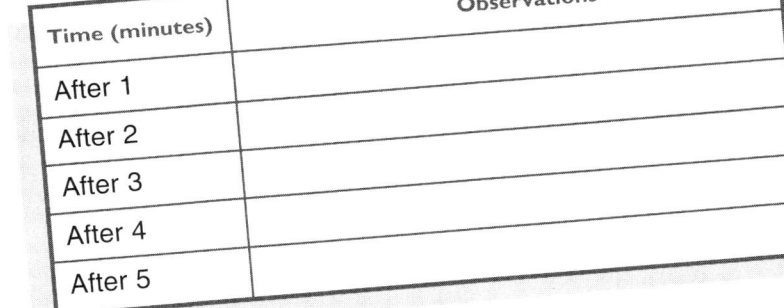

⚠ Clean up any spills immediately.

5. Look at the prediction you made and the reason you gave for your prediction.

   If something different happened from what you expected, discuss what the reasons might be.

   a) How accurate was your prediction?
   b) Did anything happen that you were not expecting? If so, what?
   c) Does your reason make good sense?
   d) If necessary, rewrite your reason to include any new information or ideas you have. Base your explanation on the evidence you have.

6. Allow your water and soil mixture to settle for 5 minutes.

   Make observations every minute during this time.

   a) Record everything you observe. An example of a data table is given below. You can change it to suit your needs.

Time (minutes)	Observations
After 1	
After 2	
After 3	
After 4	
After 5	

7. Stir your soil and water mixture with the drinking straw, then let it settle.

   Observe what is happening to the soil particles in the water. Note if all particles are behaving in the same way.

   a) Record your observations.
   b) How can you explain what you observe?

Evidence for Ideas
Consider Evidence
Seek Alternatives
Collect & Review
Collect & Review
Evidence for Ideas

Investigating Earth Systems

Teacher Commentary

5. Students revisit their predictions and reflect on how their observations compare to what they expected to occur.

    a) Student responses will vary.

    b) Student responses will vary.

    c) Evaluating whether or not their reason makes sense will be difficult for students. They may assume that, if it makes sense to them, it must make sense to others. Encourage students to think about their results, and their interpretation of the results, from the perspective of an outsider.

Teaching Tip
Revisiting Predictions
When students make predictions based on their reasoning, they have a "stake in," or "a sense of ownership" about what happens. If their prediction appears to be correct, this reinforces their reasoning. If their prediction does not appear to be correct, and they are surprised, then they will think again about their reasoning and why it was flawed. This can be a key learning moment for students. You need to help them see the importance of this. Allow them to adjust their ideas and repeat the experiment to confirm any new idea.

Understanding the Importance of Evidence
The word "evidence" is introduced here. Most students will be familiar with the term but probably in the context of criminal investigations. It might be useful to review this important scientific term using, as an example, the way evidence is used in criminal investigations. Students need to appreciate the idea that, in science, valid conclusions depend upon reliable and verifiable evidence.

6. Continue to look at student journals.

    a) This part of the investigation happens slowly. At first, it may seem that almost nothing is happening. It is important that students continue to observe the soil in water, recording all events as they happen. The sample data table should get them started. Draw their attention to the example data table as a good way of doing this. However, be prepared to accept alternative forms of recording, such as a series of labeled drawings. Later, you can discuss the relative merits of recording data in different ways.

7. Students will be familiar with some of the effects of stirring solids in water. They will know that this speeds up the process of dissolving substances like sugar and salt in water. Some of the solid particles in soil, such as salts and other minerals, may well become dissolved, but many of these will be invisible. The main effect of stirring is that it allows small solid particles to break free and move on their own. This helps particles to settle as separate layers.

Be aware of the students' conceptions about density. Many of your students will probably not have a working concept of "density", as such, but will be able to observe that some of the soil particles appear to be "lighter" than others, therefore, they end up at the top of the cup. Students could use a magnifier to help them identify these as particles of organic material. It is interesting to note that it is low mass particles (such as sand and silt), rather than low density material, like the organic matter in the bottle, that causes wind erosion problems for farmers.

a) Check to make sure that students record their observations in their journals. Emphasize the positive aspects of their work, but point out an area that could be improved as well. Too little or too much critical feedback will not be helpful. Let students know what you see, e.g., "I see good organization, but I would like to see more observations" or, "I see good ideas, but I am not sure what you are responding to...is this a prediction?"

b) Student explanations will, optimally, address mass and volume–even if they do not use proper scientific terminology. For example, a good answer will include the following: "light objects float, while heavy objects sink" or, "large objects sink while small objects float."

Teacher Commentary

NOTES

Investigation 2: Separating Soil by Settling

8. In your small group, discuss what you think will happen to this mixture if you allow it to settle for a much longer time.

    Based on what you see happening now, predict what the soil inside the cup will look like in 24 hours.

    a) Write down your prediction.

    b) Write down the reason for your prediction.

    Be sure to mark your cups so others know what is in them.

9. Label your cup in a way that you can identify it among others.

    Find a safe place to store your water and soil mixture undisturbed for the next 24 hours.

10. After 24 hours examine your soil and water mixture carefully.

    Observe how it has changed.

    a) How many kinds of materials can you now observe in the soil?

    b) What do you think these are?

    c) How do you think you could find out what the different materials are?

    d) How accurate was your prediction?

11. Design a good way to show your findings to others. Share your findings with the class.

    Record your information in a way that you can add more to it later. The following shows one way to record your findings:

What is Found in Soil	
Finding	**Evidence**
Soil is composed of many things.	We found stones, bits of...

Investigating Earth Systems

Teacher Commentary

8. Remind students to make a prediction about what they think will happen to the sample after 24 hours. Be available to student groups to help them to discuss and think about what will happen. Ask them to think of any experience they may have had with muddy water – are mud puddles clear after one day if they are not walked in?

   a) Sample predictions include: "The water will be perfectly clear because all of the fine sediment from the soil will settle out in 24 hours." Or, "The water will be clearer than it is now but not perfectly clear because more sediment will settle out in 24 hours, but not all of it."

   b) Check to make sure that students have recorded a reason for their prediction.

Teaching Tip
Making Another Prediction
By now your students should be more familiar with making a prediction based on careful reasoning. If necessary, remind them that the better the reasoning is, the more likely the prediction will be correct. Here, students are asked to make a prediction about what will happen to their soil in water cups over a longer period of time. It is important that students are given sufficient time to observe their cups settling before they make their predictions. It is just as important that students discuss their predictions about the cups (and their reasons for their predictions) with each other. Many students might be more comfortable making a drawing of their prediction.

9. Students will need to be able to find their soil samples the next day. To ensure that this happens in an orderly fashion, have students mark the names of all group members on their cup. Write names on tape or use a permanent marker. Find a secure location to store students' samples. If necessary, collect all the cups and place in a box for overnight storage.

10. Overnight, the soil particles will have separated into more easily observable layers. Organic material will have gathered toward the top of the water, though some may have become saturated and sunk below the surface. Other layers might include rock, sand, silt and clay particles. The more important thing for your students to see is that soil is composed of different materials and that water can help to separate them and make them observable. At this stage, they do not necessarily need to know that this is the result of density. However, it very important that they do not confuse this layering with the way in which soil is layered in the ground (which represents the order in which the materials have been deposited over time).

   Allow time at the beginning of class for students to revisit the experiment. After (carefully) retrieving their cups from the previous class period, instruct students to answer question 10 in their journals. They will use these answers as the basis for a presentation to their classmates.

Investigating Soil – Investigation 2

a) Students should see at least three, and probably more, different kinds of materials in layers in the cups. Gravel will be seen at the bottom, sand will be found above the gravel, and mud will be seen above the sand. Organic matter may also be seen floating in the water.

b) Good answers will include one or more of the following: plant material, leaves, blades of grass, twigs, small rocks, pebbles, sand, and mud or clay.

c) Sample answer: "Take the material out of the cup and study it under a microscope or magnifying glass."

d) This refers back to Step 8. Answers will vary. Encourage students to write more than a one-word answer such as "very" by asking them to explain their answer.

11. This is an opportunity to alert students to the fact that scientists disseminate their findings to others. The key part of this is that they should show their findings in a form that others can see and understand. To assist with this, ask students read about this inquiry process on the **Blackline Master** of **Inquiry Processes**. The results from all presentations can be summarized in the form of a table, as shown on S9.

Teacher Commentary

NOTES

Investigating Soil

INVESTIGATING SOIL

As You Read...
Think about:
1. What makes objects fall toward the center of the Earth?
2. Why do large, heavy objects fall faster than small, light ones?

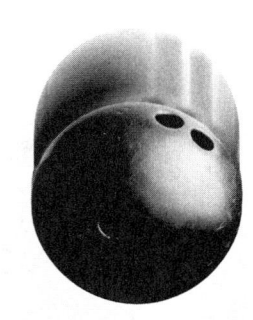

Digging Deeper

Gravity and Friction

The Earth's gravity pulls all things toward the center of the Earth. The soil materials fell to the bottom of the cup because of the pull of gravity. You know from experience that the speed of a falling object depends on the weight and the size of the object. A bowling ball falls much faster than an air-filled balloon. They are about the same size but the bowling ball is much heavier. A marble falls more slowly than a bowling ball. They are both made of the same type of material, but the bowling ball is much larger, and therefore much heavier. This probably seems very natural to you. However, it is not that simple to explain why. If you dropped those objects on the Moon, they would all fall at the same speed! The difference has to do with a force that keeps one object from moving against another object. This force is called friction. In this investigation the friction was between the air or water and the falling particles. There is no air on the Moon, so there would be no friction to hold back falling objects.

The soil in your sample probably was made up of differently sized particles. They may have been as large as sand grains or pebbles, or as small as fine mineral particles. The larger particles tended to fall faster, so they reached the bottom of the cup first. It took a long time, however, for the dust to settle in the air and for the muddy water to become clear again. Also, it took longer for the particles to settle in water than in air. That is because the friction is greater in water than in air.

Evidence for Ideas

S 10
Investigating Earth Systems

Teacher Commentary

Digging Deeper

Assessment Opportunity
Before students read the **Digging Deeper** section on gravity and friction, you might ask them to record their ideas about separating soil by settling using the following questions. Note that the answers provided are for you, not for grading purposes, as this assessment is intended to prepare students for the reading.

1. Soil fell more slowly through water than through air because _____.

   Soil fell more slowly through the water than through air because the friction that acts against the particles is much greater in water than in air.

2. Suppose you repeated Step 4 of the investigation with a tube that was 2 m tall and filled with water. Make at least two predictions about how the results might be different. What would you see? Why?

   If a longer settling tube is used, the soil would take more time to settle to the bottom (because the tube is longer), and there would be greater separation of sizes of material in the bottom of the tube (because frictional forces would have a greater opportunity to operate on the particles in the soil).

3. Draw a picture of a particle of soil that is falling through water. Label the forces that are acting on the particle as it falls.

   The diagram should show an arrow for gravity pointing down and an arrow for friction pointing up.

Assign the **Digging Deeper** reading section and **As You Read** questions to students. Discuss the answers as a class.

Digging Deeper explains how two fundamental forces (gravitational force and frictional force) affect how particles settle through fluids. Students have an intuitive understanding of gravity and friction, but their conceptual understanding is not likely to be very deep at the middle school level. The **Background Information** for this investigation will help you prepare to explain these concepts to students and to respond to students' questions.

As You Read...
1. Gravity makes things fall toward the center of the Earth.

2. Because they are less affected by the friction of the air or water.

Investigating Soil

Investigation 2: Separating Soil by Settling

Review and Reflect

Review

Evidence for Ideas

1. Can you answer the key question after completing this investigation? Explain how you were able to separate soil.
2. What kinds of materials did you find in your soil sample?
3. What is the difference between the way soil particles fall through water and through air?
4. Name two or more things that you learned about soil in this investigation.

Reflect

5. Do you think all soil is the same everywhere? If not, why not?
6. Would you expect all types of soil to behave in the same way when poured in air and water? If not, why not?
7. What additional questions about soil came up during this investigation that you could investigate?

Thinking about the Earth System

8. In which part of the Earth system is each of the following located?
   a) soil
   b) air
   c) water
9. What evidence did you find in this investigation that would connect soil to the biosphere?
10. Write all the connections that you discovered in this investigation to connect soil to the geosphere, hydrosphere, atmosphere, and biosphere. You can record this information on your *Earth System Connection* sheet.

Thinking about Scientific Inquiry

11. How is evidence used in science?
12. How can making a prediction be useful in your investigations?
13. Why is it important to give the reason for your prediction?

Teacher Commentary

Review and Reflect

The answers provided below are for you, the teacher. It is not expected that your students will answer with the same level of sophistication. Use your knowledge of the students as well as the standards set by your school district to decide on what answers you will accept. In student answers look for evidence of an understanding of the processes involved, as well as for any misconceptions that still remain. Encourage students to express their ideas clearly, and to use correct science terminology where appropriate.

Review

1. Yes. Soil was separated using air and water. By mixing soil with air and water, larger particles were separated from smaller particles.

2. A good answer will include several of the following: minerals, rocks, fine particles of silt and clay, sand, and plant remains.

3. Of course, the most obvious difference in soil movement through air and water is the speed at which the particles fall: soil falls through air faster than water. Students are likely to use the material from the **Digging Deeper** reading section to say that soil falls through air faster than through water because there is less friction between soil particles and air than there is between soil particles and water. Keep in mind, however, that the material at the end of the **Digging Deeper** reading section is in error in this regard; see the **Background Information** section in the front of this investigation. Another difference is that larger objects in the sample tended to settle first.

4. A good answer may include at least two of the following: soil can be separated by size, using both air and water, heavy things settle first because of the effect of friction, there are more things in soil than seen in Investigation 1, water is a good tool for separating soil. Accept any other reasonable answers.

Reflect

It is very important that your students are given adequate time to **Review and Reflect** on what they have done and understood in this investigation. Ensure that all students think about and discuss the questions listed here. Be on the lookout for any misunderstandings and, where necessary, help students to clarify their ideas.

5. Based only on their observations of the soil in their classroom, students should know by now that soil is far from uniform. It varies between locations and even within the same location. Students should make the connection that climate and bedrock can produce different types of soil, and that separation by air and water can separate soil components.

6. "Yes" and "No." A "yes" answer is acceptable if the student explains that gravity is the same all over the Earth and, therefore, all soils are subject to the

Investigating Soil – Investigation 2

same force and will behave in the same way. A more likely response is "no." In this case, justification hinges on students' understanding that soils are different because some have more minerals or fine particles like silt, while others have bigger particles like sand and rocks. Some will have a bigger layer of minerals, for example, or perhaps a thicker layer of sand.

7. Students may propose that they investigate soils in their local area (this is an activity they will do later–see Investigation 4: Examining Core Samples of Soil). Accept all reasonable questions.

Thinking about the Earth System

It is very important that students begin to relate what they are studying to the wider idea of the Earth System. This is a complex and largely inferred set of concepts that students cannot easily understand from direct observation. Remember, the goal is that students will have a working understanding of the Earth System by the time they complete the eighth grade. Although it can be taught as a piece of information, true understanding is largely dependent upon comprehending how numerous specific Earth science concepts connect with the idea of the Earth as a system. Be sure to spend some time helping students to make what connections they can between the focus of their investigations and this wider aspect. A **Blackline Master** (*Earth System Connection* sheet) is available in this Teacher's Edition. Students can use this to record connections that they make as they complete each investigation.

8. a) Soil is typically thought of as being primarily part of the geosphere; however students have now seen that soil is also affected by the atmosphere and hydrosphere, and can interact with both. Earthworms eat components of the soil and there are components of the biosphere in soil (as plant and animal remains). Thus, any answer can be correct if explained sufficiently.

   b) Air is not only in the atmosphere but also in the geosphere (as space between soil, mineral and rock particles) and in the hydrosphere. In addition, many living things breathe air for their oxygen, and plants take in carbon dioxide for photosynthesis.

   c) Water can be found in the atmosphere (as vapor), in living organisms, in soil and rock, and in all components of the hydrosphere.

9. Components of the biosphere were isolated from the rest of the soil in the separation activities. Plant matter was found as one of the layers in the soil.

10. Students can relate any aspects of soil to the Earth's systems. Water affects soil by moving and separating it. Soil can contain water. Elements of the biosphere get broken down and become part of the geosphere. The geosphere is utilized by elements of the biosphere, like plants, worms, burrowing animals, etc. Extremely small particles of soil might be found in the air, particularly on a windy day. The wind can break down rocks, helping to produce soil.

Teacher Commentary

Thinking about Scientific Inquiry

Science as inquiry is a theme that runs through all investigations. Students will need many investigative experiences to grasp the many processes and skills involved with scientific inquiry. This can be taught as a piece of information, but for a solid understanding, students need considerable firsthand experience in doing it for themselves. Students are given many opportunities to think about the connections between their investigations and inquiry processes. Remind students to examine their copy of the **Blackline Master** on **Inquiry Processes** when answering the following questions.

11. Just as in a courtroom, "allegations" have to be tested. The results of the test are the evidence that a scientist uses to determine whether the allegation is true. In science an allegation is called a hypothesis and evidence is proof that the hypothesis is true (and should be kept) or not true (and, thus, should be "released").

12. A prediction is useful because it proposes an answer to a question, and it can be tested to see whether it is correct.

13. If you do not give a reason for your prediction, it is no more than a guess. You may want to distinguish between a guess and a prediction with your students. The main reason not to guess is that it is a waste of time. If your question is important, it is time consuming, expensive, and wasteful to experiment "blindly."

Assessment Tool

Review and Reflect Journal–Entry Evaluation Sheet

Depending upon whether you have students complete the work individually or within a group, the **Review and Reflect** portion of each investigation can be used to provide information about individual or collective understandings about the concepts and **Inquiry Processes** explored in the investigation. Whatever choice you make, this evaluation sheet provides you with a few general criteria for assessing content and thoroughness of student work. Adapt and modify the sheet to meet your needs. Consider involving students in selecting and modifying the criteria for evaluating their end of investigation reflections.

Teacher Review

Use this section to reflect on and review the investigation. Keep in mind that your notes here are likely to be especially helpful when you teach this investigation again. Questions listed here are examples only.

Student Achievement

What evidence do you have that all students have met the science content objectives?

Are there any students who need more help in reaching these objectives? If so, how can you provide this?

What evidence do you have that all students have demonstrated their understanding of the inquiry processes?

Which of these inquiry objectives do your students need to improve upon in future investigations?

What evidence do the journal entries contain about what your students learned from this investigation?

Planning

How well did this investigation fit into your class time?

What changes can you make to improve your planning next time?

Guiding and Facilitating Learning

How well did you focus and support inquiry while interacting with students?

What changes can you make to improve classroom management for the next investigation or the next time you teach this investigation?

Teacher Commentary

How successful were you in encouraging all students to participate fully in science learning? _____

_____

How did you encourage and model the skills values, and attitudes of scientific inquiry? _____

_____

How did you nurture collaboration among students? _____

_____

Materials and Resources

What challenges did you encounter obtaining or using materials and/or resources needed for the activity? _____

_____

What changes can you make to better obtain and better manage materials and resources next time? _____

_____

Student Evaluation

Describe how you evaluated student progress. What worked well? What needs to be improved? _____

_____

How will you adapt your evaluation methods for next time? _____

_____

Describe how you guided students in self-assessment. _____

_____

Self Evaluation

How would you rate your teaching of this investigation? _____

_____

What advice would you give to a colleague who is planning to teach this investigation? _____

_____

NOTES

Teacher Commentary

INVESTIGATION 3: SEPARATING SOIL BY SIEVING
Background Information

1. Sieves and Sieving

Sedimentologists, soil scientists, and environmental scientists and engineers often use sieving to study the size distribution of the particles of minerals and rocks in soils. Sieving is the companion to settling as a way of measuring the distribution of particle sizes. The idea is simple. Make a stack of sieves, arranging them so that the size of the holes decreases downward in the stack. Put a pan at the bottom and a lid on the top. Shake the sieves with a special shaking machine for a time long enough for the particles to find their way down through the sieves. Afterward, the "catch" on each sieve is weighed, and the percentage, by weight, of the sample lying in each size interval is calculated. Many standard statistical techniques can then be used to analyze the basic data. The sieves are usually mounted in circular metal frames, and are designed to nestle tightly one in another. Sieves with openings finer than about 0.05 mm are seldom used, both because of the expense and difficulty of manufacturing such sieves, and because the ease with which particles pass through sieves decreases sharply in such fine sizes. Settling, and other technologically more sophisticated techniques, are used instead for the finest particle sizes.

Archaeologists, and also some paleontologists, use sieves in a different way: to separate out large artifacts or fossils from the bulk of fine sediment or soil material that contains the objects of interest.

2. Materials Found in Soil

Because of the enormous range in soil types, it is difficult to generalize about soil composition. All soils consist of just a few main kinds of materials: particles of minerals and rocks, living plants and animals, and decaying plant material. Most soils consist mainly of particles of minerals and rocks, but organic matter, living and dead, is an essential component of most soils.

There are only a few very common kinds of inorganic materials in most soils. Such materials vary greatly in particle size, from large boulders to colloidal particles (that is, particles that are so small that when they are in water their properties lie between particulate suspensions and true chemical solutions). In gravelly soils, most of the gravel consists of pieces of rock of various kinds. Gravel-size pieces of vein quartz, which have a milky white look, are particularly common. The sand-size particles in soils consist mostly of quartz and, in smaller concentrations, potassium feldspars and fragments of fine-grained rock types, like slate, chert, or volcanic rocks. The finest mineral particles in soils, with sizes ranging from several microns (a micron is a thousandth of a millimeter) to small fractions of a micron, are clay minerals and oxides of iron and aluminum.

Clay minerals are a group of silicate minerals with composition and structure related to common mica. They are produced by weathering of a great variety of silicate minerals in the bedrock from which the soils are derived. Several hydrous oxides of iron and aluminum are also formed by weathering of the minerals in the parent rock of the soil. The presence of iron oxides is what colors many soils red, orange, or yellow. Bauxite, the principal ore of aluminum, is soil that is particularly rich in aluminum oxides. Most bauxites are old soils, formed under an ancient warm and humid global climate.

Plants and animals that are large enough to be seen without a microscope, like earthworms or insects, are the ones you are likely to notice in a soil, but tiny single-celled organisms, mostly bacteria, algae, and fungi, are far more abundant. These microorganisms exist in most soils in enormous numbers. They are the agents that cause breakdown of dead organic matter. Dead organic matter, mostly plant material, is variably common in soils. In natural soils, the abundance of plant material is greatest very near the soil surface, and increases sharply downward. The activity of microorganisms breaks down the plant material to a very stable residue of humus, a dark, friable, and fine-grained material. Because of its high capacity for water, its porous structure, and its content of plant nutrients, it forms an excellent medium for plant root growth. Ultimately, even the humus is broken down, releasing carbon dioxide and water back into the soil environment.

The *Investigating Earth Systems* www.agiweb.org/ies web site also contains a variety of links to web sites that will help you deepen your understanding of content and prepare you to teach this investigation.

Teacher Commentary

Investigation Overview
Students deepen their understanding of soil that they will need to design a garden by sieving soil to separate soil into its component parts. Students follow a set procedure for separating soil into coarse particles, medium particles, fine particles, and floating particles. Unlike Investigation 2, where students separated soil by settling, students are now able to examine and identify the components when they have completed the separation process. Text describes the most common inorganic and organic materials found in soil.

Goals and Objectives
As a result of this investigation, students will develop a better understanding of the material found in soil and will improve their ability to follow laboratory protocols, use tools to conduct inquiry, and make observations.

Science Content Objectives
Students will collect evidence that:
1. Most soils contain many kinds of material.
2. Items in soil can be physically separated by a variety of methods.
3. Soil is a mixture of organic and inorganic matter.
4. Inorganic material in soil includes rock and mineral particles of various sizes.

Inquiry Process Skills
Students will:
1. Make observations using the senses.
2. Collect and review data using tools.
3. Use evidence to develop ideas.
4. Communicate observations and findings to others.

Connections to Standards and Benchmarks
In this investigation, students will separate soil into three sizes of material and into organic and organic material. These observations will start them on the road to understanding the National Science Education Standards and American Association for the Advancement of Science (AAAS) Benchmarks shown below.

NSES Link
- Soil consists of weathered rocks and decomposed organic material from dead plants, animals, and bacteria.

AAAS Link
- Although weathered rock is the basic component of soil, the composition and texture of soil and its fertility and resistance to erosion are greatly influenced by plant roots and debris, bacteria, fungi, worms, insects, rodents and other organisms.

Preparation and Materials Needed

Preparation

The same soil samples from Investigation 2 can be used in this investigation. The soil must be dry in order for it to pass through the various sieves and strainers that will be used to separate it into various sizes of material. If the soil has not yet dried, put the samples on aluminum foil and bake it in an oven at 200° F for a few hours.

Each group will need a plastic strainer (for coarse particles), a kitchen sieve (for medium particles), and three bowls. This works as an inexpensive alternative to purchasing a set of soil screen sieves for separating soil into various sizes of material. Have soil samples and materials available for students at the start of class.

The procedure students must follow will seem complicated, but if followed carefully, yields four different sizes of material for students to observe.

You may wish to save the soil samples for later use in Investigation 5, Part 2.

Materials
- soil sample (taken from your local area)
- large piece of white poster board
- 4 squares of white poster board (about 10 cm square)
- 3 large mixing bowls
- plastic strainer, with about 2-mm diameter holes
- kitchen sieve, with about 0.5-mm diameter holes
- large plastic cups
- plastic spoon
- hand lens

- **Optional: light microscope or binocular microscope for observing soil material**

Teacher Commentary

Making Connections ...with Mathematics

You may wish to have students extend this investigation by measuring the amounts of each size of material found in their soil sample and calculating relative mass percentages. In order to do this, students will need to have access to a mass balance. You can have them figure out how to measure mass, record the data, and calculate relative mass percents, but the basic procedure is as follows:

1. Measure and record the mass of the entire dry sample.
2. Measure and record the mass of each piece of poster board.
3. Complete the separation of soil into components (Steps 1 – 6).
4. Measure and record the mass of each sample (plus poster board).
5. Subtract the mass of the poster board from the mass of poster board plus sample. This gives you the mass of each sample.
6. Divide the mass of each sample by the total mass of the sample and multiply the result by 100. This gives you the relative weight percent of each component.

If the samples were collected in different areas, have students compare results and use this as an opportunity to explore how soils vary from place to place.

The steps above are also provided as **Blackline Master** *Soil* 3.1 (Measuring and Calculating Mass of Materials Found in Soil).

Investigating Soil

INVESTIGATING SOIL

Investigation 3:
Separating Soil by Sieving

Key Question
Before you begin, think again about the key question from the last investigation.

How can soil be separated?

In the last investigation, you used the method of settling to separate the different materials in a soil sample. You probably also thought of other ways to separate the materials.
Share your thinking with others in your group and with your class. Make a list that combines everyone's ideas about other methods for separating soil.

Materials Needed

For this investigation your group will need:

- soil sample (from your local area)
- large piece of white poster board
- 4 squares of white poster board (about 10 cm (4") square)
- 3 large mixing bowls
- plastic strainer, with about 2-mm diameter holes
- kitchen sieve, with about 0.5-mm diameter holes
- large plastic cups
- plastic spoon
- hand lens

Investigate
1. Spread the soil sample onto a piece of poster board to dry.
   When it is completely dry, break up any lumps by pressing gently on them with your thumb.

Teacher Commentary

Key Question

The **Key Question** is a brief warm-up activity that draws out students' ideas about the topic explored in the investigation.

Write the **Key Question** on the board or overhead transparency. Encourage students to give detailed answers. How do the pictures show more refined techniques than in the previous investigation? The pictures on pages S12 and S13 offer additional clues. Tell students to record their ideas in a new journal entry.

Discuss students' ideas. Ask for a volunteer to record responses on the board or overhead projector. This allows you to circulate among the students, encouraging them to copy the notes in an organized way.

Student Conceptions

Students' ideas about separating soil should be more advanced after the first two investigations. Students may respond to the **Key Question** with answers that at this point should be fairly obvious: soil can be separated with water and air.

Answer for the Teacher Only

The **Background Information** for this investigation and the previous investigations provides a detailed discussion of settling and sieving as two methods that can be used to separate soil. The *Investigating Earth Systems* www.agiweb.org/ies web site also contains a variety of links to web sites that will help you deepen your understanding of these methods.

> ### Assessment Tool
> **Key–Question Evaluation Sheet**
> The **Key–Question Evaluation Sheet** will help students to understand and internalize basic expectations for the warm-up activity.

Investigate

Teaching Suggestions and Sample Answers

Introducing the Investigation

You will need most of the class period to complete the investigation. Tell students that they will be responsible for completing Steps 1 – 6 of the investigation during class. If you assign roles to students, choose responsible students for "cleanup" duty. Before beginning, you may want to cover desks and tables with newspaper.

Remind students that, as always, laboratory protocol requires them to wear safety goggles. In addition, they should always wash their hands after handling soil.

> **Teaching Tip**
> The steps that the students need to follow may at first seem complicated, especially to students who experience difficulty reading, or ESL students. Pair these students with good readers. Suggest that the students read the steps together and then have the student with reading difficulties draw small diagrams to explain the procedure prior to the class beginning the investigation. Followed carefully, the steps are very straightforward. These students may then wish to share their diagrams with others as they proceed through the investigation.

1. Ideally, the soil should be dry at the start of the investigation. It is important to break up the lumps in the soil. Advise students that, although this takes a little extra time, the results will be significantly better.

Teacher Commentary

NOTES

Investigation 3: Separating Soil by Sieving

Conduct Investigations

2. Put the plastic strainer over one of the large bowls.

   Pour your soil sample into the strainer.

   Shake the strainer to make the finer part of the soil pass through.

   If you see any more lumps of soil in the strainer, break them up gently and shake again.

⚠ Shake the strainer gently.

Conduct Investigations

3. Pour the material from the strainer into a large plastic cup.

   Fill the cup with warm water.

   Stir the water with a spoon for about 10 s.

   Skim off any floating material with the spoon, and put it on a small square of poster board to dry.

Collect & Review

4. Stir the water again and wait two or three seconds.

   Pour the water off. Stop just before you lose the coarse material at the bottom of the cup.

   Repeat this a few times, until there is almost no fine cloudy material in the water.

   Dump or scrape the remaining material onto another small square of poster board to dry.

⚠ Label all samples that are left to dry so that others know what they are.

Teacher Commentary

2. Observe students: Check to make sure that students are recording their actions. Let them know that their journals are records of what they do in class. This is **not** the same as copying the steps from their *Investigating Soil* book. Suggest ways of organizing their entries. For example, a step could have two entries: "action" and "observation."

   As always, careful classroom management is the first priority in maintaining a safe classroom. You may wish to demonstrate how to go about shaking the strainer so that soil is not spilled.

3. Skimming the floating material off the top is tedious work, but very necessary for the success of the investigation. Tell students to take the extra time to remove as much of the floating material as possible (although it may be impossible to get all of it).

   You may want students to place floating soil particles on a paper towel before transferring to poster board. Poster board holds up well under moist conditions, but it should not be soaked.

   Remind students to label their poster board squares so that others know what the material is.

4. The point of this step is to remove the pebbles and other large, heavy items from the soil. The supernatant (the liquid poured off the top) will not be used further. You may want to have students do this over the sink. Instead of pouring the muddy water down the drain, you may want to have students pour it into a bucket or other container. This avoids clogging up your sinks.

   In the event that water spills get on the floor, wipe them up immediately.

Investigating Soil

INVESTIGATING SOIL

5. Put the kitchen sieve over the second bowl, and pour the dry soil that passed through the strainer into the sieve.

Repeat steps 3 and 4.

Combine any floating material you caught in this step with the floating material you caught earlier.

6. Now you have separated your soil sample into four parts: coarse particles, medium-sized particles, fine particles, and floating particles.

After each part has dried, examine it with the hand lens under strong light. Try to identify the materials in each part. It helps to examine the material both when it is wet and when it is dry.

It will be much harder to identify the materials in the finest part than in the coarser parts. Even soil scientists, using specialized equipment, have trouble with the very finest materials in soils.

a) Record your observations on a data sheet. Note the size of the particles, the shape of particles, the color of the particles, and what you think the particles are made of.

b) On another sheet of paper, write down detailed descriptions of each item in your data table. Describe everything about the materials you can think of. Each detail, no matter how small, might turn out to be very important when you try to interpret your observations.

Inquiry
Making Observations

Careful observations and descriptions of materials and processes are the basis for all good science. Scientists first describe what they see, as carefully as they can. Only then do they try to form theories and interpret their data.

Clean up spills immediately.

As You Read...
Think about:
1. Why is it important to have methods of separating soil samples?
2. What things are you likely to find in soil?

Digging Deeper

Materials Found in Soil

Soil scientists separate soil using a stack of several sieves. The sieves have holes with slightly different sizes. The coarsest sieve is at the top of the stack, and the finest sieve is at the bottom. You used a similar but simpler method in this investigation. This method of separating the soil materials gives you the best chance for looking at the different kinds of materials. You can think of it as a "divide and conquer" method!

Teacher Commentary

5. In Steps 2 – 4, coarse particles, and much of the floating material, were separated from the soil mixture. This means that both medium and fine materials passed through the strainer. In Step 5, fine particles will be separated first from the material left in the large strainer (or colander). This is easily accomplished by passing it through a finer strainer (a kitchen sieve or similar device with holes 1 mm or less in diameter). Fine particles move through the fine strainer. With the material left in the fine strainer, students repeat the previous steps. First they place the mix in a large cup and skim off the floating materials (adding to their collection of twigs and organic matter), then they pour off the liquid. Basically, the students are rinsing the remaining (medium-sized) particles of floating matter and fine matter.

   Students should place the four types of soil particles on poster board squares and label them (floating, coarse, medium, and fine).

6. Make sure that students have access to and use a hand lens to observe the various components of the soil sample. Remind students that their journals are an important tool for recording observations and evidence about the nature of the soil sample. Point out the description of **Inquiry Processes** in the margin on page S14 (**Inquiry – Making Observations**). Help students to understand that careful descriptions are very important because they provide the basis for making sound interpretations.

   a) Pay attention to how students organize information in their journals. The "data sheet" referred to in the student books can be something they create in their journals. It should describe their observations concisely and thoroughly.

   b) Check to make sure that students are describing the materials they see in each of the four components, and that their records are clearly labeled and descriptive.

Teaching Tip
If you have a binocular microscope or light microscope, allow students to observe the material with these. Students may enjoy looking at their soil through a light microscope (especially the smaller particles). Have them draw pictures of what they see, recording the level of magnification used.

Assessment Opportunity
Checking for Misconceptions
Have students read the following paragraph and ask them whether or not they agree or disagree and explain why:
Soil is made up of two things: mud and organic matter (such as plants and roots). When a clump of soil is put into water, the clay falls to the bottom of the container and the organic matter floats on top.

Answer for the Teacher Only

By this point of the investigation, students should understand that soil is a mixture of material, including rock fragments, sand, silt, clay, and a variety of organic material. They should have observed differences in the sizes of material during their three investigations into soil and be using this information to develop an understanding of soil as a mixture.

Digging Deeper

The reading section explains the materials found in soil. The first paragraph describes how soil scientists use screen sieves to separate soil into various sizes of material.

As You Read...

1. It is important to have methods for separating soil because soil components have different properties, relating to plant growth and drainage for example.

2. A good answer will include several of the following: rock particles, mineral particles (mainly quartz, feldspar, and clay minerals), decaying plant material (sticks, leaves, humus), living plants and animals, water and air in pore spaces.

Teaching Tip

The **Background Information** at the start of this investigation in the Teacher's Edition will help you to prepare for a lecture or discussion about the materials found in soil. You may want students to be able to see the things that the reading refers to. Keep enough sets of the four types of soil particles so students can look at them while they read. You may want to incorporate the soil samples into a discussion of the second **As You Read** question (What things are you likely to find in soil?).

For additional teaching suggestions on how to present the content of this **Digging Deeper** reading section to your students, please refer to the *IES* web site.

Teacher Commentary

NOTES

Investigating Soil

Investigation 3: Separating Soil by Sieving

Most soils contain many kinds of material. All soils consist mainly of two kinds of material: particles of minerals and rocks, and organic matter. Organic matter is any matter that is or was once living.

Your soil is likely to have several kinds of rock and mineral particles. A few kinds are very common. Many other kinds are sometimes common, but usually are not. The three most common kinds are quartz particles, feldspar particles, and small pieces of rock. Your sample is very likely to have a lot of at least one of these three kinds of particles.

Quartz particles have irregular shapes. They look gray and glassy. Their surfaces are often stained brown or orange, because they are coated with rust. Feldspar particles are usually white or cream-colored. Their surfaces are often flat, at least partly, rather than irregular. There are many kinds of rock particles. You can tell them apart from the mineral particles because rocks are made of many different particles of minerals, all stuck tightly together.

The finest part of your soil is probably mostly very small flakes of clay. They are too small for you to see even with a hand lens. Sandy soils are loose and easy to dig. Soils with a lot of clay are harder to dig. Some plants prefer sandy soils, and others soils with more clay.

Investigating Earth Systems

S 15

Teacher Commentary

The material on this page explains the inorganic material found in soil. If your students have studied rocks and minerals prior to this module, this provides an excellent opportunity to revisit what they have learned about basic rock and mineral identification and properties.

About the Photo
The photograph of a soil sample reveals rock fragments, sand, finer material, and organic matter.

Assessment Opportunity
You may wish to select questions from the **As You Read** section to use as quizzes, rephrasing the questions into multiple choice or "true/false" formats. This provides assessment information about student understanding and serves as a motivational tool to ensure that students complete the reading assignment and comprehend the main ideas.

INVESTIGATING SOIL

> Most soils have lots of organic matter. Some of the organic matter is in the form of living things. The ones you might see in your sample are large, like earthworms or insects. There are also many very tiny plants and animals, called microorganisms. There are many more of these, but you can't see them without a microscope. In a typical soil, there are millions of them in every cubic centimeter!
>
> Most soils are also rich in decaying plants. If the plant has decayed only slightly, you can usually recognize scraps of leaves, roots, and seeds. When the plant has decayed more, it turns into a soft, fine, dark material called humus. Humus is very important in soils. New plants can easily put their roots into humus. It is also good at holding water for later use by growing plants.

Review and Reflect

Review

1. Which kinds of materials in your sample came from plants?
2. Which kinds of materials came from animals?
3. Which kinds of materials formed in the soil, and which were present before the soil began to form?
4. Which kinds of material might have dropped into the soil from the sky?

Reflect

5. Why was it important for your study of soil, to be able to separate a sample of soil?

Thinking about Scientific Inquiry

6. What type of evidence did you collect in this investigation?
7. How did the use of tools in this investigation help you answer the key question?

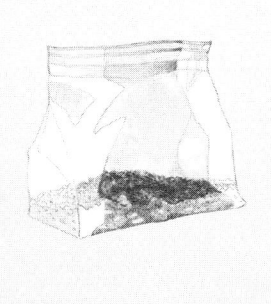

Teacher Commentary

> **Making Connections ...with Biology**
>
> The last two paragraphs of the reading explain the organic matter found in soil. This provides an excellent opportunity to make connections with biology and the study of plants and animals found in soil.

Review and Reflect

Review

Be sure to give students adequate time to review what they have done and understood in this investigation. Ensure that all students think about and discuss the questions listed here. Be alert to any misunderstandings and, where necessary, help students to clarify their ideas.

1. Most of the floating material probably came from plants. This material is usually dark in color, and has a wide range of sizes, from identifiable pieces of leaves, stem, seeds, etc., to very fine material (humus) that is in an advanced state of decay.

2. Most of the animal material floated as well. Dead insects are most likely to be seen.

3. In general, plant and animal material most likely formed on or in the soil. Some of the rock and mineral particles are likely to have been present before the soil began to form. Other particles, especially clay minerals, are likely to have formed in the soil. These last materials are almost always very fine grained.

4. Materials that may have dropped from the sky include: insects, plant seeds, and also very fine-grained mineral material, mainly particles of clay minerals, which might have been blown in by the wind as dust.

Reflect

5. Students will be planning a garden. They need to know what is in the soil so that they can make informed decisions about what they should plant. Accept any reasonable answer.

Thinking about Scientific Inquiry

In this investigation, your students have been using scientific tools: sieves, hand lenses, and perhaps, microscopes. Have your students refer to the fourth **Inquiry Process** listed on the **Blackline Master** of **Inquiry Processes** "Collect and review data using tools." Ask them to look carefully at the explanation given for this **Inquiry Process**. In addition, help students to see how they used other inquiry processes, such as using evidence to develop ideas.

6. Students collected evidence about the sizes of material found in a soil sample and the types of material found in a soil sample.

7. The strainers and sieves were tools used to answer the **Key Question**. The strainer removed fine and medium-sized particles (and some floating material). The material that remained in the strainer (coarse particles and floating particles) were separated from each other using water. Next, the medium particles were separated from the fine particles with an even finer strainer. Floating materials were removed from the medium-sized particles using water.

Teacher Commentary

NOTES

Teacher Review

Use this section to reflect on and review the investigation. Keep in mind that your notes here are likely to be especially helpful when you teach this investigation again. Questions listed here are examples only.

Student Achievement

What evidence do you have that all students have met the science content objectives?
_____

Are there any students who need more help in reaching these objectives? If so, how can you provide this? _____
_____

What evidence do you have that all students have demonstrated their understanding of the inquiry processes? _____
_____

Which of these inquiry objectives do your students need to improve upon in future investigations? _____
_____

What evidence do the journal entries contain about what your students learned from this investigation? _____
_____

Planning

How well did this investigation fit into your class time? _____
_____

What changes can you make to improve your planning next time? _____
_____

Guiding and Facilitating Learning

How well did you focus and support inquiry while interacting with students?
_____

What changes can you make to improve classroom management for the next investigation or the next time you teach this investigation? _____
_____

Investigating Earth Systems – Investigating Soil

Teacher Commentary

How successful were you in encouraging all students to participate fully in science learning? _____

_____

How did you encourage and model the skills values, and attitudes of scientific inquiry? _____

_____

How did you nurture collaboration among students? _____

_____

Materials and Resources

What challenges did you encounter obtaining or using materials and/or resources needed for the activity? _____

_____

What changes can you make to better obtain and better manage materials and resources next time? _____

_____

Student Evaluation

Describe how you evaluated student progress. What worked well? What needs to be improved? _____

_____

How will you adapt your evaluation methods for next time? _____

_____

Describe how you guided students in self-assessment. _____

_____

Self Evaluation

How would you rate your teaching of this investigation? _____

_____

What advice would you give to a colleague who is planning to teach this investigation? _____

_____

NOTES

Teacher Commentary

INVESTIGATION 4: EXAMINING CORE SAMPLES OF SOIL
Background Information

1. Topographic maps
Topographic maps are a two-dimensional representation of a three-dimensional land surface. They use lines (or, more precisely, curves) of equal elevation, called contour lines, to show the elevation of the land. A topographic map shows the relief (variation in elevation) of the land surface. Contour lines can be thought of as boundaries that separate areas above that are higher in elevation from areas below that are lower in elevation. The contour interval of a contour map is the difference in elevation between adjacent contour lines.

Here are some characteristics of topographic maps:

- The closer together the contour lines in some area of the map, the steeper the slope of the land in that area.
- Contour lines can never cross one another, although two or more can merge together where there is a vertical cliff.
- On most topographic maps, every fifth contour line on a map is darker and its elevation is always marked at various points along it.

Topographic maps also show many features on the surface, including water bodies, vegetation, roads, buildings, political boundaries, and place names. Experienced users of contour maps are able to visualize the Earth's surface topography (the "lay of the land") in their minds just by examining a topographic map of an area.

2. Soil Profiles
Most soils vary greatly in their composition and structure with depth below the land surface. That is to be expected, because soils develop from the top down, under the influence of weathering processes, plant growth, and the activity of animals. The degree or extent of soil development decreases with depth, until at depths typically of a meter or two (but greater in humid tropical climates), the effects of soil development are minimal. The thickness of a soil depends not only on the intensity of the processes that act to produce soils but also on the local slope of the land. In areas with steep slopes, the downslope movements of newly developed soil material (which can be very fast, as in landslides, or very slow, as in soil creep) tend to remove soil as it develops, whereas in areas with very gentle slopes, soils develop in place without any lateral movement.

Soil scientists use the term soil profile for a vertical section through a soil. The study of soil profiles is an important focus of research by soil scientists, because it helps in understanding the nature of soil formation processes. Knowledge of the vertical structure of soils has practical importance as well, because it provides some ability to predict the nature of the subsurface from observations that are limited to the surface.

The characteristics of the soil change continuously with depth, but soil scientists conventionally recognize a number of distinctive layers, called horizons. In soils that support abundant plant growth, the uppermost layer of the soil is a thin layer of plant litter, which grades down into plant material, called humus, that is in a more advanced state of decay. This uppermost horizon is called the H horizon. Below that, and present in all soils, is the A horizon. The A horizon, with a thickness usually in the range of a decimeter to a meter, is the main upper layer of all soils.

Investigating Soil – Investigation 4 **111**

In areas like the eastern U.S., where rainfall is abundant and the climate is humid, the A horizon experiences leaching due to net downward movement of rainwater from the surface. Chemical weathering reactions break down all but the most resistant minerals. The products of the chemical weathering in the A horizon are typically clay minerals, in the form of very fine particles, and dissolved ions.

The solid weathering products work their way downward, into the B horizon, by various processes, including physical transport and dissolution–reprecipitation, to become concentrated in the B horizon. In areas where the upper layers of the soil are rich in quartz, which is very resistant to weathering, the A horizon takes on a characteristic grayish-white appearance in its lower part, below where organic matter is abundant. The B horizon contains less weathered material, together with weathering products from the A horizon. In many places the B horizon is colored various shades of yellow, orange, or red, because of precipitation of hydrous iron oxides; the iron is derived from weathering of iron-bearing minerals higher in the soil profile.

The C horizon, below the B horizon, consists of only slightly weathered material, which grades down into the parent material of the soil. Such soils are characteristically acidic. In many soils in the central and western U.S., where rainfall is less and evaporation is more intense, there is less distinction between the A and the B horizons, because return of water to the surface by capillary rise tends to reprecipitate weathering products back into the A horizon. Also, where calcium-bearing minerals are present in the parent materials, fine-grained calcite (a calcium carbonate mineral) is common in such soils; it is formed when the dissolved calcium ions react with carbonate ions supplied from carbon dioxide from above. These soils are characteristically neutral to alkaline rather than acidic.

The *Investigating Earth Systems* www.agiweb.org/ies web site also contains a variety of links to web sites that will help you deepen your understanding of content and prepare you to teach this investigation.

Teacher Commentary

Investigation Overview

Students conduct a field study in their school grounds or nearby area to investigate how soil varies from place to place. Students begin by collecting a core sample of soil and noting the location of the sample on a map of the sampling area. Students bring the sample back to the classroom and conduct a detailed analysis of the core sample. They observe and describe the sample and then slice the sample open to conduct further observations. Students summarize and compare their analyses with those of other groups and search for patterns and relationships in the character of the soil from place to place. Students combine results to create a soil map and raise additional questions for inquiry that can help them to improve their map. Students extend their study of soil by consulting topographic maps, soil maps, and other sources of information about the nature of the land and soil in their region.

Goals and Objectives

As a result of this investigation, students will develop a better understanding about the nature and distribution of soil types around this school and improve their ability to design and conduct scientific inquiry.

Science Content Objectives
Students will collect evidence that:
1. Soil is often layered.
2. Soil layers have particular characteristics (color, texture, etc.).
3. Soil can contain both living and non-living components.
4. Soil composition varies depending on location.
5. Soil is part of the Earth's System.

Inquiry Process Skills
Students will:
1. Follow a protocol to collect core samples of soil.
2. Use map skills to record where soil samples were collected.
3. Use tools (such as magnifiers) to make observations about soil core samples.
4. Use their senses to make observations about soil core samples.
5. Make a record of the nature of the soil core sample using words, tables, and pictures.
6. Organize findings into a display.
7. Find evidence to support conclusions.
8. Communicate data and conclusions to others.
9. Compare conclusions between groups.
10. Generate new questions to investigate about soil samples and their sources.

Connections to Standards and Benchmarks
In this investigation, students will collect and analyze a core sample of soil and compare it to other samples. These observations will start them on the road to understanding the National Science Education Standards shown below.

NSES Links
- Soils are often found in layers, with each having a different chemical composition and texture.
- Students should develop the ability to identify their questions with scientific ideas, concepts, and qualitative relationships that guide investigation. (NSES, page 145).

Teacher Commentary

Preparation and Materials Needed

Preparation

Parts 1 and 2:
In this investigation, your students will collect their own soil core samples from various points in the school grounds. Your students will then use the samples to investigate how the soil may be different at different points of your school grounds. If your school grounds do not have suitable places to take soil samples, try to find an alternative spot close to the school. This could be a local park, a grass verge alongside a street, or anywhere where soil sampling will not be a problem.

Your students will go outside to take soil samples. They will need to be well-prepared to work in groups for this event. Make sure each student in a group knows what his or her responsibility is in taking samples, recording where the sample is from, preserving the sample for analysis back in the classroom, and so on.

It is important that you have a map of the area your students are going to sample. Your school administration might have a site plan that includes the school grounds. If not, you could create your own, or have your class work together to do this. There are some obvious mathematics connections here (measurement, scale, etc.) and it might be worth talking to your students' mathematics teachers about possible collaboration. The map does not need to be too detailed but it will help if it is reasonably accurate and has interesting landmarks on it (such as hills, dips, grassland, wooded areas, and landscaped areas).

It is strongly recommended that you try this method of taking core samples yourself prior to having students do it. Core sampling will also be easier if one end of the PVC pipe has been sharpened or beveled with a file to provide a cutting edge. Getting the core out of the pipe can be difficult if the soil is particularly sticky. Spraying the inside of the pipe with non-stick cooking spray will help the core to release more easily. If you have scientific soil core equipment available, you should substitute that for the PVC pipe. Following a set procedure in science is an important skill that your students should practice and understand.

It is difficult to take core samples if the ground is hard. In addition, different areas of the country have different types of soil, with some being more dense than others. You might consider watering the spots where soil core samples are going to be taken well ahead of sampling time. You will need to judge if it is safe for your students to take core samples. If in doubt, take the core samples yourself while your students watch. It is important that they witness the core sampling process.

Students need to take hammers, wood blocks, PVC pipes, rulers, plastic wrap, maps of the sampling area, wooden dowels, marking pens, and paper with them to the sampling site. You might have them bring a plastic bag to class to carry materials and supplies into the field. Obtain permission, if necessary, to collect soil samples from areas near your school.

Investigating Soil – Investigation 4

Part 3:
Make a transparency of the sampling area map so that it will be available to students when they compare their results in Step 7 of Part 3 of the investigation.

Part 4:
Part 4 calls for outside resources and will require some advance preparation on your part if you wish to reduce the time students spend searching for information. For example, it will be helpful to investigate sources of information on soil types in your region. The *IES* web site will be helpful www.agiweb.org/ies. The Student Book also refers to topographic maps used by geologists and other professionals in their work. As they are fairly complex to interpret, your students are likely to need your help using them. The maps for your area can usually be obtained from the office of the United States Geological Survey (USGS) for your state or region and ordered through the Internet (see www.agiweb.org/ies for helpful hints about how to obtain local topographic maps). As an alternative, you could use a good road map that shows elevations, rivers, and land use. These maps often provide less detail than a local topographic map, but students should be able to make some reasonable connections between soil characteristics and origins. Maps from the American Horticultural Society for your region may also be very useful.

Materials
- map of sampling area
- piece of 2.5 cm (1") heavy duty PVC pipe about 25 cm (10") long
- metric ruler
- wooden block
- hammer
- garden or work glove
- piece of wooden dowel 30 cm (12") long that fits inside PVC pipe
- plastic wrap
- masking tape and marker (or other method of labeling core sample)
- colored pencils
- plastic knife
- hand lens
- tweezers or tongue depressor

Teacher Commentary

NOTES

Investigating Soil

Investigation 4:
Examining Core Samples of Soil

Key Question
Before you begin, first think about this key question.

Is all soil the same?

You have been looking at one soil sample. Think about how this sample would compare with samples from different places. Share your ideas with others in your class.

Investigate

Part 1: Collecting the Soil Samples

1. To investigate differences in soils, it is important to collect all samples using the same procedure.

    Once you have selected a site, use the procedure on the following pages to collect each sample.

Materials Needed

In this investigation your group will need:

- map of sampling area
- piece of 2.5 cm (1") heavy duty PVC pipe about 25 cm (10") long
- wooden block
- hammer
- garden or work glove
- piece of wooden dowel 30 cm (12") long that fits inside PVC pipe
- large sheet of white poster board or paper
- plastic wrap
- colored pencils
- plastic knife
- hand lens
- tweezers or tongue depressor

Teacher Commentary

Key Question

The **Key Question** is a brief warm-up activity that draws out students' ideas about the topic explored in the investigation. This **Key Question** is designed to find out what your students think about whether or not soil is all the same, and to set the stage for inquiry.

Write the **Key Question** on the board or overhead transparency. Encourage students to think about what they learned in earlier investigations and readings and to support their answers with evidence and reasons. Tell students to record their ideas in a new journal entry.

Discuss students' ideas. Ask for a volunteer to record responses on the board or overhead projector. This allows you to circulate among the students, encouraging them to copy the notes in an organized way.

Student Conceptions

By this point in the investigation (particularly if you have had students bring in soil samples from home) students will have begun to note that soil differs from place to place, and have thus begun to develop some understanding about variations in soil. Students who have traveled to different parts of the country may know that the color of soil can be remarkably different in different regions. Some students will understand that since organic matter is part of soil, soil in regions that are devoid of plant matter is likely to be different from soil in regions that have lots of plants. Finally, students who have gone camping or backpacking and spent the night sleeping in a tent may recall that soil can be thin, rocky, and/or hard in some places and soft in other places.

Answer for the Teacher Only

Soil varies remarkably from place to place. A perfectly appropriate answer to the question "Is all soil the same?" is "Yes and No." Evidence for both positions comes from previous investigations and readings. Most soils have the same components: rocks, minerals, organic material. Soils form, in part, from bedrock. The list of similarities could be quite long. So, too, could the list of differences. Relative amounts of rocks, minerals and organic material vary from location to location. Different bedrock gives rise to different types of soil.

Assessment Tool
Key–Question Evaluation Sheet
The **Key–Question Evaluation Sheet** will help students to understand and internalize basic expectations for the warm-up activity.

Investigate
Teaching Suggestions and Sample Answers

Part 1: Collecting the Soil Samples

1. Tell students that they are going to collect soil samples from areas around the school. Show them the equipment they will use and explain how the samples will be collected. Students will need to take hammers, wood blocks, PVC pipes, rulers, plastic wrap, maps of the sampling area, and wooden dowels with them to the sampling site. Since students will be comparing their samples at a later time, it is important that all samples be collected using the same procedure.

   Advise students to read the procedure on the following pages before they go into the field. Make sure that they understand basic protocols for safety and field work before you leave the classroom.

> **Teaching Tip**
> This may be your students' first investigation in the field. It is important that students realize that scientific inquiry is not limited to laboratories. Much of Earth science research takes place in the field. Sampling soil, rocks, and water is a major part of this research. Sometimes tests are done in the field, but more often samples are taken back to laboratories for detailed study and analysis. It is important that your students appreciate this dimension of science as inquiry, and that **Inquiry Processes** are used in field studies as well as in laboratories.

> **Assessment Tools**
>
> **Investigation Journal–Entry Evaluation Sheet**
> This sheet will help students to learn the basic expectations for journal entries that feature the write-up of investigations. It provides a variety of criteria that both you and students can use to ensure that student work meets the highest possible standards and expectations. Adapt this sheet so that it is appropriate for your classroom. You may also wish to make modifications to develop a sheet specific to a given investigation.
>
> **Journal–Entry Checklist**
> This checklist provides you and your students with a guide for quickly checking the quality and completeness of journal entries.

Teacher Commentary

NOTES

Investigating Soil

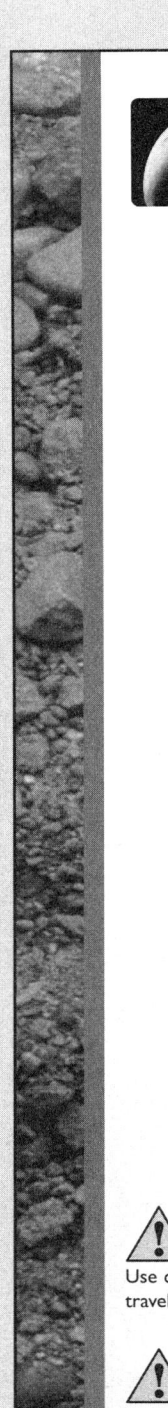

INVESTIGATING SOIL

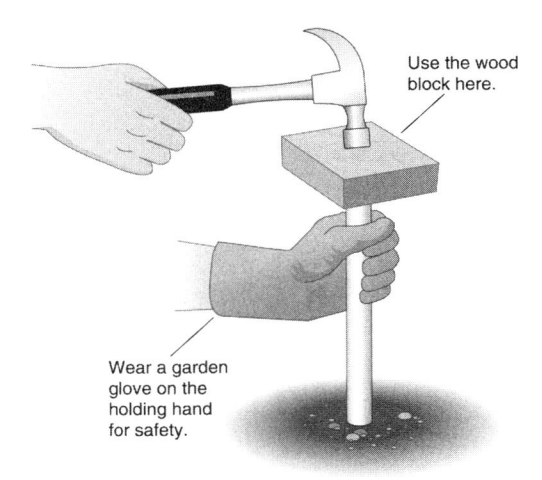

Use the wood block here.

Wear a garden glove on the holding hand for safety.

1 Stick the sharp end of a PVC pipe into the ground. Set a wooden block on top. Be sure to wear a garden or work glove on the hand holding the pipe.

Collect & Review

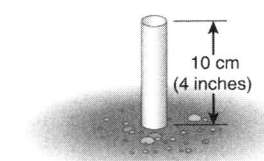

Conduct Investigation

10 cm (4 inches)

2 Hammer the pipe into the ground leaving 10 cm (4") sticking above the surface.

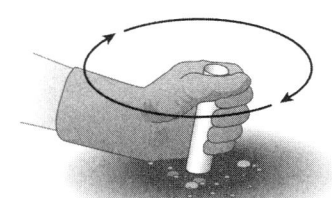

3 Grasp and move the pipe gently in a circular motion to loosen it from the ground. If the pipe is stuck, use gentle side-taps with the hammer to free it. Carefully pull the pipe out of the ground. If the soil does not remain inside the pipe, repeat the procedure somewhere else nearby.

⚠ Use caution in traveling to the site.

⚠ Collect only in areas where you have permission.

Teacher Commentary

1. Advise students who have not had much practice with a hammer to hit the wood straight on, not at an angle. Consider demonstrating the safe procedure for taking a sample before allowing students to collect samples on their own. You will need to judge if it is safe for your students to take core samples. If in doubt, take the core samples yourself while your students watch. It is important that they witness the core sampling process.

2. Students may not be able to drive the PVC pipe into some soils that are too hard. In some cases this is because there is a rock in that spot. By moving a few feet away it may be easier to collect a sample. Tell students not to try and force the pipe into the ground. In cases where local soil is extremely hard, you may need to prepare spots in advance by adding water and letting it percolate through the soil overnight. Remind students to measure the PVC pipe so that 10 cm (4") of the pipe remains sticking above the ground. Students may make the measurement in the field, or they may wish to make the measurement in the classroom

3. Students should not be concerned if it is difficult to pull the pipe out of the ground. Have them gently twist it and, if necessary, tap gently on the side of the pipe while twisting.

Investigating Soil

Investigation 4: Examining Core Samples of Soil

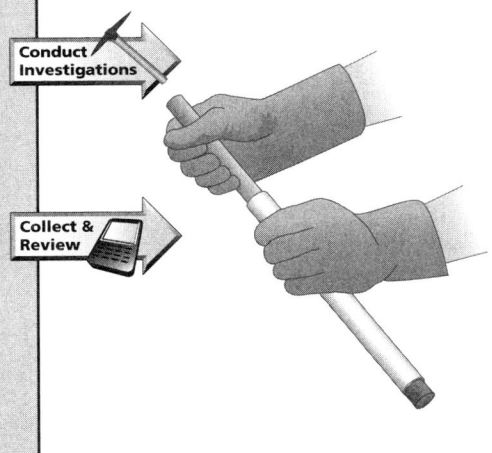

Conduct Investigations

Collect & Review

4. Insert a dowel into the open end of the pipe. Carefully push the core of soil out onto a flat surface covered with a sheet of plastic wrap.

5. If the soil core is very firm, gently tap the dowel with the hammer to release it. Make sure that the soil core stays together as much as possible.

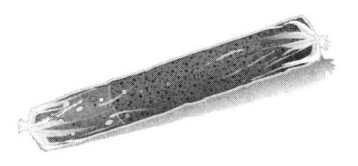

6. Carefully wrap the core in the plastic wrap. If you roll the core up in several layers of the plastic wrap, it is more likely to stay intact while you carry it back to the classroom.

2. On a map of the sampling area, record the exact location where the soil sample was taken.

⚠ Collect soil samples with adult supervision only.

⚠ If living organisms are present inform your teacher.

S 19

Investigating Earth Systems

124 Investigating Earth Systems

Teacher Commentary

4 This step should be done slowly. Turning the dowel gently as it pushes the soil core out of the PVC pipe helps the process.

5 It is helpful to have the plastic wrap ready on the ground when the core is extruded or removed.

6 Students should label their core samples, perhaps with a piece of masking tape attached to the plastic wrap. They should label the top end of the sample to indicate which end is up.

2. It is very important that students record exactly where each soil core comes from on the map. Make sure that each core is labeled with its origin. In comparing differences between the cores, students will be able to connect soil characteristics with soil origin.

   Pay special attention to where your students get their core samples. Vegetation and land elevation might indicate differences in the underlying soil. Many school grounds have been landscaped, especially in open, grassy areas. The soil there might be different from areas that seem to have been undisturbed for many years (wooded areas with old trees, or dips in the ground where water may run off in heavy rainfall). Look for variety.

   ### Making Connections ...with Mathematics

   Suggest ways that students can use mathematics to mark the location of their soil samples. If you used graph paper to make a map of the sampling area, this is easier; however, it is not the only way. First they need to find the scale. They can do this by measuring the length of something on their map. By comparing the real length of the object with the "map length" of the object, you can approximate distances and get coordinates for the sampling site.

Investigating Soil

INVESTIGATING SOIL

Part 2: Analyzing the Soil Core Samples

1. Prepare your area in the classroom to analyze the soil core samples.

   Select the materials you will need.
   Spread a sheet of poster board on your table area.

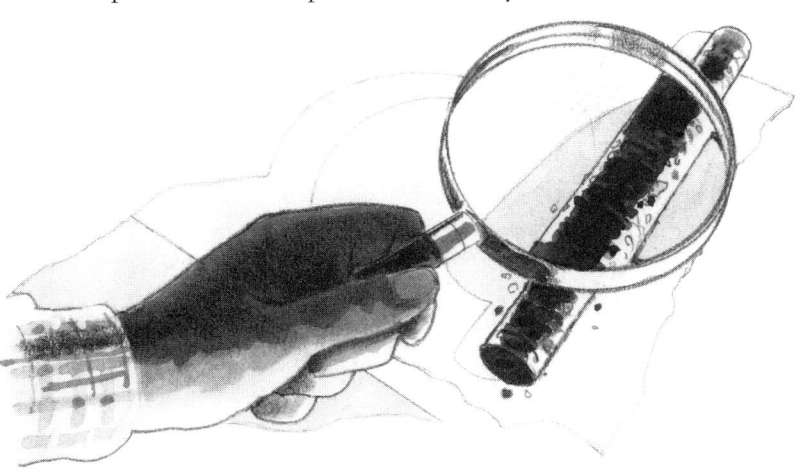

2. Lay the core on the poster board, and unwrap it without breaking it. Observe it carefully.

   Using just your eyes, note any differences in the core from the top to the bottom. Look for color changes, different types of material, and any living things.

   a) Record any interesting things you observe.

   b) On the poster board, next to the core, draw a picture of the core. Use colored pencils to make the drawing as realistic as possible. Label the parts of the core you observed.

3. Use a plastic knife to make a long cut down the center of the core. Look at the inside.

   a) Record any interesting materials or items you uncover. Add them to your diagram.

4. Let the core dry out thoroughly.

   Gently break any lumps of soil with your thumb.

   With the plastic knife, spread the core material apart sideways until it is about 5 cm wide. Now you can study the materials in the core in more detail.

⚠️ Break lumps of soil gently or, if they are too hard, discard them. Clean up spills immediately.

Teacher Commentary

Part 2: Analyzing the Soil Core Samples

1. Cover desks with newspaper before beginning. Remind students about basic safety procedures for handling soil, including the use of disposable latex gloves, goggles, lab aprons, and washing hands thoroughly after handling soil.

2. Observe students as they investigate. Encourage them to make observations in their journals. Look at their journal entries. Check to see that they are including dates and headings for each entry. Check for thoroughness of observations. Note that the word "data" is used freely in the Student Book. You may want to check that your students have a good understanding of this scientific term.

> **Teaching Tip**
> Poster board can be substituted with butcher paper.

a) Your students have had some experience analyzing soil in Investigation 1. This time they need to be more detailed in their approach. Here, they are not just looking to see what different materials are in the soil, but where they are placed within a core sample, and how much of each is in the core. As groups work, observe what they do and remind them that they will be comparing core samples with other groups.

b) Make sure that students draw and label a picture of their sample before taking the sample apart. Students may not realize that graphic representations of objects are also a valid part of scientific inquiry. They need to appreciate that drawings can be vital evidence in some circumstances. An alternative to drawing is to take a photograph of each core sample, preferably with a camera that yields instantaneous results. If students do this, they should place a metric ruler beside each sample to indicate the scale.

Again, the key element here is that students should display their results in a form others can see and understand. This is especially important if students are making direct comparisons. Your students should be able to appreciate the need for a clear display from a scientific point of view. Following the same procedure will ensure this.

> **Teaching Tip**
> Core samples must be dry before starting Step 4. This would be a good stopping point in the investigation. Have your students leave the samples unwrapped so that the samples can dry out overnight.

4. If core samples were taken during the previous class period they should be dry enough for use. It is not necessary that the samples be completely dry. However, the sample should be dry enough so that its components can be easily separated.

Investigating Soil

Investigation 4: Examining Core Samples of Soil

Collect & Review

5. Use a hand lens to observe the different kinds of materials. Use the knowledge you gained in the last investigation to help you identify the materials.

   a) Record your observations. A data table like the one shown below can be helpful. You can make up your own table and use categories to suit your sample.

Material group	Type of material(s)	Color(s) of materials	Particle size	Name of material
1	stones	red/yellow	pea size	rock chips
2				
3				
4				
5				

Evidence for Ideas

6. In your group, examine your data.

   Consider and discuss the following questions:

   - Which materials in your core sample contain living (or once living) things? Which materials in your core sample have probably not come from living things? How can you tell? Which materials are you uncertain about?

   - Which materials tend to be near the top of the soil sample? Which tend to be near the bottom?

   - Which materials seem to be all through the core sample?

   - How could measurements help you to describe your sample?

   - What could you do to make measurements of the samples?

Explore Questions

   a) Record any further data you uncovered in your discussion.

Teacher Commentary

5. Observe students as they record their observations. Check to see that they have a well-organized system for recording their data. The data table shown is just one option for recording data. You could suggest to your students that they devise their own method. If so, the method should allow students to log findings in an organized and reliable way. The data table shown can also be adapted to suit students' findings. The data table is available as **Blackline Master** *Soil*, 4.1 (Data from Analysis of Soil-Core Sample).

> ### Making Connections ...with Mathematics
>
> Encourage students to use any measurements they can take to supplement their data. These can include the width of any layers in the core, the size of any rocks or pebbles, the mass of each core, and, after breaking down the core into parts, the mass of different materials.

6. Encourage students to work together on this step. They should discuss their answers. It is important that students realize that these questions are for discussion. They are not necessarily questions for which there is a correct or specific answer. Point out to them that once data have been collected, the next step is to form and ask the kinds of questions that will help explain the data. The questions help in interpreting the evidence collected. In this sense, questions are used both to start an investigation, and to develop explanations for the observations later on.

Sample Student Responses
- Living things found in the sample may include worms, insects, insect exoskeletons, and plant debris. Non-living items in the sample will likely include rocks, minerals, clay, sand and silt. Students may note that, like in their settling experiment, living (or formerly living) things tend to be near the top of the sample.
- Answers in group discussions will typically vary.
- Answers in group discussions will typically vary.
- Students could use measurements to quantify and compare layers.
- Students could use rulers to take measurements.

   a) Although the questions initiated are for discussion purposes, stress that you expect a reasonable response recorded in the students' journals.

> ### Assessment Opportunity
> Select and have students complete one of the following questions to check for understanding of soil maps and how soil maps are useful:
> - Our soil map will be useful for our final investigation because it _____.
> - One thing that would improve my soil map would be to find out _____.
>
> *(continued)*

Investigating Soil – Investigation 4

Answers for the Teacher Only

Our soil map will be useful for our final investigation because it will show us what the soil is like in different places and help us plan the garden.

One thing that would improve my soil map would be to find out what nutrients are in the soil in different places.

Teacher Commentary

NOTES

Investigating Soil

INVESTIGATING SOIL

Inquiry
Studying Patterns and Relationships

Finding your own evidence does not always provide a complete explanation to a scientific question. Scientists look for evidence other scientists have collected. They then look for explanations by studying patterns and relationships within the evidence.

⚠️ Have your plan approved by your teacher before you begin.

7. To make a soil map of your sampling area, you will need to present your findings and compare them to the findings of other groups in your class.

   Mark the location of each soil sample on one map of the area.

   Look for patterns and relationships between the soil samples and the areas where they were found. Consider the following:
   - What plants and animals are found at each location?
   - Is the location mostly high ground or mostly low ground?
   - What evidence of water is there? Is the area dry, moist, or wet?

   a) Think about similarities and differences among the soil samples. Try to group similar soil samples together on your map. Look for patterns of soil types, and record any patterns you find.

Part 3: Extending the Investigation

1. Your map will probably not be as complete as you would like. There may be places where you do not have enough data. There may be samples that raise further questions to investigate.

   Discuss what additional evidence you need to obtain to improve your soil map. Think about other questions you would like to investigate.

   a) Record questions you would like to investigate.

2. Using these questions, put together a reasonable plan for gathering the information you need.

   a) Record your plan for further investigation.

3. With the approval of your teacher, carry out your investigation.

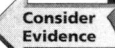

Teacher Commentary

7. After groups have followed the research procedures, the whole class will be ready to compare results. You will need to organize the sharing session so that everyone can make clear data comparisons. One way of sharing findings is to have each group take turns in presenting. Alternatives to this might include having pairs of groups compare first, then have a pair join with another pair, and so on. Another way might be to have groups make displays. Once again, students should understand that the "dissemination of findings" is a key part of science as inquiry.

   It may be useful to make transparencies of a sheet of graph paper and the sampling site map. Show this on the overhead projector. Give each group a transparency sheet with graph paper copied on it. Now, each group can figure out the coordinates of their sampling site, and place them accurately on the master map, shown on the projector.

   Point out the note about **Inquiry Processes** on page S22 in the margin of the Student Book ("Studying Patterns and Relationships"). This step in the investigation is a vital link between the "lab" work students have completed and the outside world in which the core samples were taken. In studying the map, students may be able to infer the reasons for some of the differences in the sample evidence. Encourage everyone to look for possible connections between soil characteristics and origin.

   a) Discuss this step as a class after all groups have presented their findings. Help students see patterns by making a chart. Along the side, you may wish to list several main soil types. Along the top, you may list plants/animals, high/low ground, dry/moist/wet. Students should refer to their maps as the class discusses the data.

Part 3: Extending the Investigation

This extension provides students with practice in obtaining information from a variety of sources. Whether or not you do the extension depends upon the level of your students' understanding and the availability of materials and resources.

> ### Teaching Tip
> The Internet is a wonderful resource for information. The *IES* web site provides an excellent starting point for research at www.agiweb.org/ies. Some of your students may be quite proficient in exploring the Internet. Encourage students to search using key words and a good search engine.

1. Students may not yet realize that investigations, as well as providing some answers, also tend to generate new questions for investigation. They may need your help in seeing that this is part of the nature of scientific inquiry. Encourage students to think about their previous experiments. How sure are they that what they have found is "true"? Have they collected enough information to grow a successful garden? What further information do they need? Perhaps students are

starting to think about new questions. They must be questions that could be answered through inquiry. In essence, this means it must be possible to make predictions from them. In addition, their question must be answerable, using only those materials available in the classroom.

a) Questions may have arisen from the previous discussion of soil maps. Student questions should be ambitious but answerable:
- Do soils at low ground level have more rocks and pebbles? Do they have less?
- Is there a relationship between the quantity of living material in a soil sample and the quantity of plants growing at the site?

2. a) All students should record their plan in their journals.

3. Student plans should be based on questions that can be solved through inquiry. It must be possible to predict an answer to the question. Observe students as they investigate their questions.

Teacher Commentary

NOTES

Investigating Soil

Investigation 4: Examining Core Samples of Soil

Part 4: Using Other Resources

1. If possible, obtain professionally produced maps to provide you with additional information.

   Use print and electronic resources to gather additional information about soil and soil types.

   Consult a local gardening expert or civil engineer to obtain information about your local soil.

   a) Record any additional information you gather. Be sure to include the source from which you obtained the information.

Conduct Investigations

Evidence for Ideas

Digging Deeper

Topographic Maps

There are many kinds of maps. Almost all of the U.S. is covered by topographic maps. Topographic maps show you land elevations such as hills and valleys. The hills and valleys are marked with a series of curves called contour lines. A contour line connects all the points on the land surface that are at the same elevation.

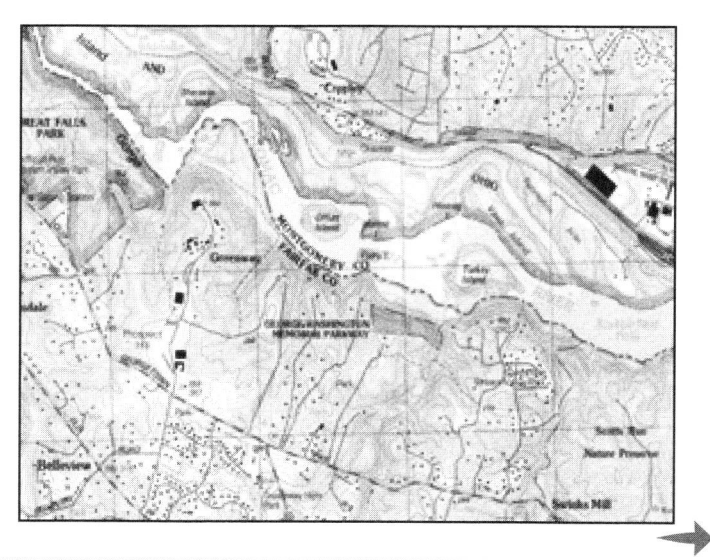

As You Read...
Think about:
What does a topographic map show you?

Teacher Commentary

Part 4: Using Other Resources

They may not be able to find a soil map for your neighborhood, city or county. Emphasize that there are many levels of research. From asking a farmer about a particular spot on his farm, to general reference information on their state's soil, all sources are open for discussion.

Topographic maps can be ordered through the United States Geological Survey. They are not free, however, and a lot of free information is available on the web. Encourage students to look at all the levels of information they can find. For example, there may be little information on soil types in their neighborhood, but there is certainly information on their state and/or region. Suggest search terms like "soil profile," "topsoil," etc.

a) Stress to the students that noting the source of information gathered is extremely important. Be sure to give students guidelines on what form of written information you want to see from them. The source name and URL is essential. Perhaps they could write a five-sentence summary of the site, or share what they learned with the rest of the class.

Digging Deeper

The reading explains topographic maps. The map shown on page S23 is part of a topographic map of northern Virginia. The Potomac River is shown in blue (as is all water on USGS topographic maps). Areas of vegetation are shown in green. Human-made structures are shown in black (homes, schools, and industrial buildings, for example). Contour lines reveal changes in the elevation of the land.

As You Read...

A topographic map shows land elevations in an area. It also shows bodies of water and boundary lines (county lines and state lines, for example).

Investigating Soil

INVESTIGATING SOIL

You can use a topographic map to see where streams and rivers flow. The area of the land that drains into a particular stream is called the watershed of that stream. Soil types usually differ quite a lot depending on where they are located in the watershed. Soils near the stream or river are usually thicker and better developed than soil in highlands.

You can also tell from a topographic map how steep the slope of the land is. The soil type often depends on the slope of the land. Soils on gently sloping land are usually thicker and better developed than soils on steep slopes.

In many areas of the United States, soil scientists have made soil maps, which show where different soil types are located at the Earth's surface. Your map is just this kind of soil map! If you can obtain a published soil map of your area, you can compare your map with the published map.

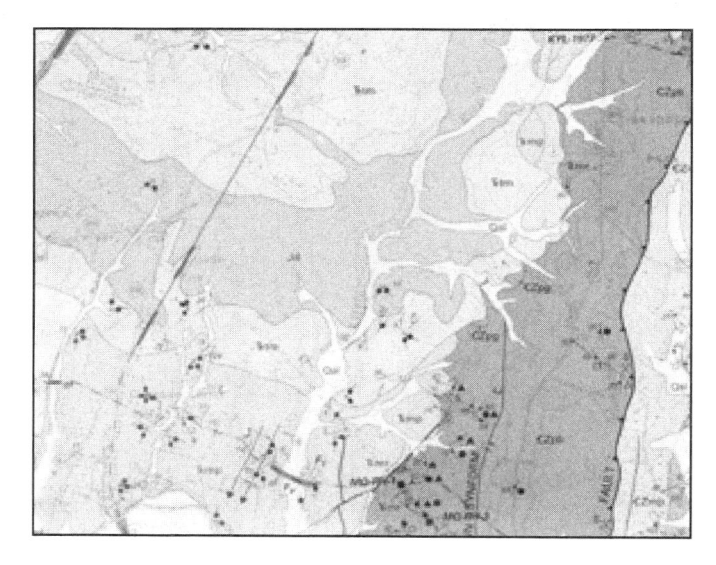

Teacher Commentary

The map on page S24 is a geologic map. It shows the rock formations and geologic structures (faults and folds) at the Earth's surface. A map key (not visible) reveals how each color represents a group of rocks of a particular age. Soil maps also represent the unique type of soil distributed over the surface of the land with colors or symbols.

Investigation 4: Examining Core Samples of Soil

Review and Reflect

Review

1. What questions about soil has this investigation answered?
2. What characteristics of soil did you use to separate your soil samples?
3. What evidence did you use to identify living and non-living components of your soil sample?
4. How could you find out if two soil samples might have been taken from the same location?

Reflect

5. What have you learned about soil by comparing soil samples taken from different areas?
6. Based on what you now know, what new questions about soil do you think would be useful to investigate?

Thinking about the Earth System

7. Write any new connections that you have made between soil and the Earth system on your *Earth System Connection* sheet for soil.

Thinking about Scientific Inquiry

8. How did you use tools to collect data in this investigation?
9. In this investigation, why was it important for all groups to display their findings using the sample procedure?

Teacher Commentary

Review and Reflect

Review
It is important that all students review carefully what they have done in this investigation, especially those events that have extended their understanding of soil.

1. Answers will vary. Students should see that soil samples varied and that these variations could be studied by mapping them.

2. Soil samples could have been separated by color, particle size, or texture. Accept all reasonable answers.

3. Evidence should include: animals in the soil sample move; living plant material is green; non-living materials do not move.

4. One way to find out if the soils are from the same area is to compare the layers. Soils from the same area should have the same color and texture. The layers should have the same characteristics and be of the same thickness.

Reflect
Students may not yet realize that investigations provide some answers and also tend to generate new questions for investigation. Students may need your help in seeing that this is part of the nature of scientific inquiry.

5. One thing that comparing soils can show is how different the samples can be from one place to another. All samples should have certain things in common (layers, for example) but the layers may look very different. Certain particles, like pebbles or sand, can be present in greater or lesser concentrations.

6. Accept all reasonable responses. Examples include: What would the soil look like if the core sample went deeper? Why do some of the samples look so different from each other? Which of the samples represent soils that would be best to grow vegetables in?

Thinking about the Earth System
Students should now be able to connect what they have been doing with all four parts of the Earth System. Remember, it is the interconnections between these systems that students must come to appreciate. Through discussion, they should be able to see that soil is located in the geosphere, water in the hydrosphere, and air in the atmosphere, and that the biosphere is represented by materials in soil that have come from living things.

7. A good answer will include one or more of the following: living things like worms or plants can be found in the geosphere. Decomposition is affected by the hydrosphere and atmosphere. The shape of the geosphere (topography) can affect other aspects of the geosphere (like soil thickness).

Thinking about Scientific Inquiry

Once again, your students need to reflect on the scientific inquiry aspects of the investigation. They may need your help to see that they have been investigating soil through a set of clear scientific processes that are generic to many investigations. In particular, make sure students realize that making a reasoned prediction at the start of an investigation is an important part of the inquiry process. Take time to discuss the inquiry procedure that your students used to compare soil samples. The method of investigating they have used will provide them with a model to use as they do other investigations.

8. Examples of tools used in this investigation include the hammer, PVC pipe, and wood block used to collect soil cores, the plastic knife (or other tool) used to look inside the soil sample, and topographic maps used to study the area.

9. It was important for all groups to use the same procedure and then share their findings, so the soil map could be more detailed and could be used for comparing findings.

Teacher Commentary

NOTES

Teacher Review

Use this section to reflect on and review the investigation. Keep in mind that your notes here are likely to be especially helpful when you teach this investigation again. Questions listed here are examples only.

Student Achievement

What evidence do you have that all students have met the science content objectives?
_____

Are there any students who need more help in reaching these objectives? If so, how can you provide this? _____
_____

What evidence do you have that all students have demonstrated their understanding of the inquiry processes? _____
_____

Which of these inquiry objectives do your students need to improve upon in future investigations? _____
_____

What evidence do the journal entries contain about what your students learned from this investigation? _____
_____

Planning

How well did this investigation fit into your class time? _____
_____

What changes can you make to improve your planning next time? _____
_____

Guiding and Facilitating Learning

How well did you focus and support inquiry while interacting with students?
_____

What changes can you make to improve classroom management for the next investigation or the next time you teach this investigation? _____
_____

Teacher Commentary

How successful were you in encouraging all students to participate fully in science learning? _____
_____

How did you encourage and model the skills values, and attitudes of scientific inquiry? _____
_____

How did you nurture collaboration among students? _____
_____

Materials and Resources

What challenges did you encounter obtaining or using materials and/or resources needed for the activity? _____
_____

What changes can you make to better obtain and better manage materials and resources next time? _____
_____

Student Evaluation

Describe how you evaluated student progress. What worked well? What needs to be improved? _____
_____

How will you adapt your evaluation methods for next time? _____
_____

Describe how you guided students in self-assessment. _____
_____

Self Evaluation

How would you rate your teaching of this investigation? _____
_____

What advice would you give to a colleague who is planning to teach this investigation? _____
_____

NOTES

Teacher Commentary

INVESTIGATION 5: WATER AND OTHER CHEMICALS IN SOIL

Background Information

1. Water in Soil

A porous medium is a material that contains empty spaces, called pore spaces, among its constituent solid materials. Soil is a porous medium. The porosity of a porous medium is defined as the ratio of the volume of pores to the total volume of the medium, in a representative sample of the medium. The ratio is usually multiplied by 100, to express it as a percentage. The porosity of very porous materials like loose sand or gravel can be as high as 30%. A typical soil might have a porosity of only a few percent, because when finer particles are present in the interstices among larger particles, the porosity is less.

One of the important aspects of porous media like soils is that fluids (air and water are the important ones in soils) can flow through them. It is always important to consider how fast, as well as in which direction, the fluid flows through the medium. The speed of flow is determined by a property of the medium called its permeability. The permeability of a porous medium is a measure of how easily a fluid can be forced through the medium by imposing a difference in fluid pressure from one place in the medium to another place. An easy way to grasp the concept of permeability is to imagine stuffing one of your home water pipes with sand just upstream of a valve or faucet. When you turn the water on, the pressure of your water system tends to force water to flow through the porous sand and out the faucet (provided that you install a screen just upstream of the faucet to keep the sand in the pipe). Very porous media tend to have correspondingly high permeabilities, but two qualifications are needed: (1) if the pores in the medium are very small, flow through them is slow, because of the increased friction; and (2) in some media, like solid rocks, some or even most of the pores are not physically connected, so the fluid cannot flow from pore to pore.

Soils, which are right at the Earth's surface, experience downward flow of water through them when it rains. The passage of surface water into the soil is called infiltration, and the continued movement of the water downward, toward the groundwater table, is called percolation. During a heavy rain, all of the pore spaces in the soil are filled with downward-flowing water, but after the rain stops, most of the water drains away, leaving the pores filled mostly with air and other soil gases. The water content of a soil immediately after such drainage is called the field capacity of the soil. This is a very important characteristic of soils for agriculture and horticulture, because it represents the water that is available for plants between rainstorms or irrigation. Coarse soils, with abundant sand and gravel, are very porous but they have relatively low field capacities. Finer soils are likely to have much higher field capacities if they contain abundant plant material, most of which has a micro-porous structure and acts much like a sponge.

2. Dissolved Constituents in Soils

Even rainwater is not pure: it contains dissolved atmospheric gases. One of the important dissolved gases is oxygen, which participates in chemical weathering by oxidizing the ferrous iron contained in many major rock and soil minerals to ferric oxides and hydroxides. Another important dissolved gas is carbon dioxide. A small percentage of the dissolved carbon dioxide reacts chemically with the water to form a weak acid called carbonic acid. Because of this carbonic acid, soils that are freshly supplied with rainwater are rather strongly acidic (as characterized by their low value of pH, considerably less than the neutral value of 7). In areas where rainfall is scarce and infrequent, the soil water tends to become more nearly of neutral pH, or even alkaline, by reactions with various mineral constituents in the soil, especially calcium carbonate minerals. The calcium ions (Ca^{2+}) released in this way is what makes some water "hard."

As soil water resides in the soil, it causes the slow dissolution of many other solid soil constituents. This releases small concentrations of various ions into the soil water. Among these, potassium ions (K^+) are an important plant nutrient. Most of the potassium released into solution this way is taken up by plants, and is then recycled when the plant dies and decays. Dissolved phosphorus, which comes from certain minerals but also from bone material, is another important plant nutrient. Still another important plant nutrient, nitrogen, comes not from minerals but in two other ways: (1) dissolved in rainwater, after having been produced from atmospheric nitrogen gas by the locally extremely high temperatures in lightning bolts, and (2) fixed from soil nitrogen gas by microorganisms called nitrogen-fixing bacteria, which live in the roots of certain plants like beans and peas (called legumes).

The *Investigating Earth Systems* www.agiweb.org/ies web site also contains a variety of links to web sites that will help you deepen your understanding of content and prepare you to teach this investigation.

Teacher Commentary

Investigation Overview

Students begin by expressing their ideas about the chemicals that may be found in soil. They go into the field to examine a soil profile by digging an 18-inch deep pit in the soil. Students predict how water will move through soil and then test their prediction by pouring water into the soil pit. In the second part of the investigation, students use their observations from the field study to design an investigation to test how water moves through various particles found in soil: plant material, sand, gravel, and clay. In Part 3 of the investigation, students conduct a second field study of soil and perform a percolation test to investigate how water moves through soil. In the last part of the investigation, students use a soil test kit to determine the concentrations of various chemicals and nutrients in soil. Students add the information from these four investigations to their soil maps from a previous investigation, which helps them to prepare for their final investigation – planning a garden.

Goals and Objectives

As a result of this investigation, students will develop an understanding of the permeability of soils by completing soil percolation tests. They will also gain knowledge about some of the chemicals in soil through the use of test kits to measure nutrients.

Science Content Objectives

Students will collect evidence that:
1. The nature of soil allows water to percolate through it.
2. Particle size affects the rate at which water percolates through soil.
3. Soil can be analyzed for nutrient content and pH levels.
4. Drainage rates and soil chemistry affect plants that live in soil.

Inquiry Process Skills

Students will:
1. Collect a sample according to a set procedure.
2. Predict an event.
3. Design and conduct an investigation.
4. Organize and present findings to others.
5. Use findings from a laboratory test to explain a field study (percolation).
6. Use evidence to verify or refute a prediction.
7. Follow a set procedure for a laboratory test.
8. Compare test results to standards.
9. Interpret test results in terms of their practical implications.

Connections to Standards and Benchmarks

In this investigation, students will conduct investigations into the nature of soil drainage and soil chemistry. These observations will start them on the road to understanding the National Science Education Standards and American Association for the Advancement of Science Benchmarks shown below.

NSES Links
- Soil consists of weathered rocks and decomposed organic material from dead plants, animals, and bacteria. Soils are often found in layers, with each having a different chemical composition and texture.
- Students should develop general abilities, such as systematic observation, making accurate measurements and identifying and controlling variables.

AAAS Links
- Although weathered rock is the basic component of soil, the composition and texture of soil and its fertility and resistance to erosion are greatly influenced by plant roots and debris, bacteria, fungi, worms, insects, rodents, and other organisms.
- At this level, students need to become more systematic and sophisticated in conducting their investigations, some of which may last for weeks or more. That means closing in on an understanding of what constitutes a good experiment. The concept of controlling variables is straightforward but achieving it in practice is difficult. Students can make some headway, however, by participating in enough experimental investigations (not to the exclusion, of course, of other kinds of investigations) and explicitly discussing how explanation relates to experimental design.

Teacher Commentary

Preparation and Materials Needed

Preparation
In this investigation your students need to be able to see how a soil sample looks from the surface down and to a depth of about 45 cm (18 in.). Note the soil profile diagram on page S27 of the Student Book. The geological term "soil profile" refers to a vertical section of soil that displays all its horizons.

Part 1: Observing a Soil Profile
This is an outdoor investigation. You will need to dig one or more pits in soil to:
a) enable students to sketch a "soil profile" or cross section of the soil;
b) pour water on the ground adjacent to the hole in order to see how water moves through soil, and
c) obtain a cubic foot of soil to bring back to your classroom as an exhibit.
This will involve finding a suitable site in, or close to, your school grounds where a hole can be dug. Dig when the soil is not too hard or soft. You will need a flat shovel, a large watering can filled with water, and a cardboard box (to transport the soil cube back to the classroom). Common sense dictates that you should fill in any pits dug in areas where students may be walking after you have left the site. The number of soil pits that you dig will depend upon the suitability of your site and the number of students in your class. (One pit may prevent all students from making observations in the time you have available outdoors.)

Take care to keep the soil cube intact. If the soil is crumbly or dry, you may not be successful in obtaining a cube of soil for display in the classroom. Consider taking digital photographs or video if you wish to have a record of the soil profile for classroom use.

Optional Materials: Students will draw a soil profile (Part 1, Step 1). Depending upon the time available outdoors, you may wish to have them make a color drawing. If so, you will need to remind them to bring colored pencils or crayons to the site. You may also wish to bring meter sticks or metric rulers to the site so that students can record the thickness of any soil layers. At the very least, students will need their journals and a pen or pencil.

In order to observe water percolating through soil (Step 3 of Part 1), pour water onto the grass in the area adjacent to the hole that you have dug. After you pour the water, take the shovel and make a new vertical slice through the wall of the hole about two inches in from the edge. This will enable students to observe and record the percolation of water down through the soil. If it has rained heavily recently and the soil is saturated, you may not be able to see the results.

Part 2: Designing and Conducting Your Own Investigation
This part of the investigation takes place in the classroom. Each group will need eight clear plastic drinking cups (four of which should have a hole in the bottom to allow water to drain) and enough plant material (peat moss will work), sand, gravel, and clay for students to do this investigation. To obtain clay, run a dry clayey soil sample

through a kitchen sieve and collect the material that passes through the sieve. (You can use the material saved from Investigation 3.)

Part 3: Conducting a Percolation Test
This is an outdoor investigation that your students can conduct on the school grounds. Students experiment with the way in which water percolates through soil by driving an empty soup can into the ground, then filling the can with water. Select a site for the percolation test in advance. It should be easy to supervise, as well. The four steps in the procedure are illustrated and explained in the Student Book. You might feel that it would be best to hammer the soup cans into the ground yourself. Use your judgment about this. Soup cans are fairly fragile. In very hard ground they may buckle or bend. Large stones or tree roots can also be a problem. It will be worth having some spare cans available. You will need to find good places to drive the cans into the ground. A pointed stick pushed into the ground will tell you where good spots are for the test. Again, the relative wetness of the soil will be important.

Part 4: Testing for Other Chemicals
You will need soil-test kits for your students. The instructions and information they contain on their packaging are usually well written and easy to understand. Review the **Background Information** provided in the instructions, paying careful attention to all safety warnings that accompany the kit.

Making Connections ...with the Professionals

It might be useful to contact the water authorities in your community to find out who conducts percolation tests (perc tests). Often, there are engineering companies that do this. You (or your students) could find this out by looking up "percolation tests" in the index of your telephone book's Yellow Pages. It would be very helpful if you could arrange for a class visit from a person who does perc tests professionally. Your students could then compare the professional method of doing the test with their "amateur" methods.

Materials
- garden shovel
- watering can
- supply of water
- 4 clear-plastic (10 oz.) cups with small hole in bottom
- 4 clear-plastic (6 oz.) cups
- cup of sand
- cup of gravel
- cup of clay
- large soup can with ends removed (extras are recommended)
- wooden block
- hammer
- garden or work gloves
- timing device
- soil-testing kit

Teacher Commentary

NOTES

Investigating Soil

INVESTIGATING SOIL

Investigation 5:
Water and Other Chemicals in Soil

Materials Needed

For this investigation your group will need:
- garden shovel
- watering can
- supply of water
- 4 clear-plastic (10 oz.) cups with small hole in bottom
- 4 clear-plastic (6 oz.) cups
- cup of sand
- cup of gravel
- cup of clay
- 2 large soup cans with ends removed
- wooden block
- hammer
- garden or work glove
- timing device
- soil-testing kit

Key Question
Before you begin, first think about this key question.

What chemicals are in soil?

By now you know that soil contains materials that come from both living and non-living things. These are the particles, big and small, that are fairly easy to see. Think about what other things, that are not so easy to see, might be in soil. Share your thinking with others in your group and with your class.

Investigate

Part 1: Observing a Soil Profile

1. As a class, go outside to the site that has been selected for you.

Teacher Commentary

Key Question

Have students read and record their answer to the **Key Question** in their journals. Write the **Key Question** on the board or overhead transparency. Tell students to write the **Key Question** in their journals and to think about and answer the questions individually. Tell them to write as much as they know and to provide as much detail as possible in their responses. Emphasize that the date and the prompt (question, heading, etc.) should be included in journal entries.

Student Conceptions

Although your students have explored soil in many ways, this is the first time they will have focused on chemicals. Students at this age are not likely to have formal conceptual understanding of the molecular theory of matter, which will impact their ability to understand and appreciate the chemistry of soil. Students may not have many ideas at all about the types of chemicals in soil, and are most likely to note things like "minerals," "iron," "pollution," and possibly "fertilizer." Students may also not regard water as a chemical. In the everyday world, people often talk about adding chemicals to water. This implies that water itself is somehow not a chemical substance. You may want to check your students' informal ideas about what a chemical is before beginning on this investigation.

Answer for the Teacher Only

Soil contains a variety of chemicals, with the exact chemical composition of soil depending upon a variety of factors. Important dissolved chemicals in soil that provide nutrients for plants include potassium, phosphorus, and nitrogen.

Investigate
Teaching Suggestions and Sample Answers
Part 1: Observing a Soil Profile

As a result of their investigations, students should understand that soil layers change with depth. Soil profiles can be widely different depending on location and underlying soil conditions. However, the general rule is that plant life is near the top, along with darker organic matter (humus). The next layers consist of lighter colored rock particle material and plant roots, larger rock particles (little or no plant life), then bedrock (this may be too deep to expose with a shovel). Compare the soil profile to this diagram. You could ask students to dig their own trenches at home, make a sketch of the soil profile and bring the sketch back to the classroom for comparison (remind students to dig only with parental approval and supervision). Comparing data like this will help to show the variety of soil conditions in your community.

1. Gather materials needed for this outdoor investigation. Make sure that students understand your expectations for safety and behavior for outdoor work before you leave the classroom. Use the following question in a class discussion to check for their understanding:

    - What question are we trying to answer in the field study today?

Investigating Soil

Investigation 5: Water and Other Chemicals in Soil

Conduct Investigations

Dig into the soil with a shovel and expose the soil to a depth of about 45 cm (18").

Observe how the soil characteristics change the deeper you dig.

If possible, dig out a 30-cm (1') cube of soil to transport back to your classroom as an exhibit.

a) Use the soil profile shown in the diagram as a guide for making a sketch of the soil you have exposed. (The diagram represents a profile of soil that is old and has not been disturbed.)

Use caution when using the shovel. After use, place it where it will not be stepped on or tripped over.

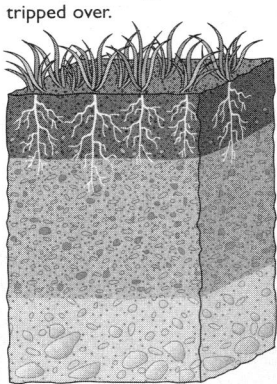

Conduct Investigations

2. Predict what you think will happen if you sprinkle a can full of water over the top of the soil you have exposed.

a) Record your prediction and the reason for your prediction.

Collect & Review

3. Gently pour all the water from a full watering can over the soil. Make it as much like rain falling as possible.

a) Where does the water seem to go in the soil?

b) How can you tell that the water is moving through the soil?

Evidence for Ideas

c) Were your predictions and your reasoning accurate? If not, how can you explain any differences?

4. Repair the site where you exposed the soil.

Take your sketches and observation notes back with you to the classroom.

When repairing the site, be sure that the hole will not be a hazard for passers-by.

Only dig in approved areas with permission of the landowner.

Investigating Earth Systems

S 27

156 Investigating Earth Systems

Teacher Commentary

1. Choose a location with soil that is not too hard. Dig a hole 18 inches deep in the soil. As you dig, try to section off a one foot by one foot square of soil for your soil cube. Try to remove the cube intact and place it into a container for transport back to the classroom.

   a) When you have finished digging and have produced a cleanly exposed face (on at least one side of the soil pit), instruct students to draw in their journals what they see in the vertical section of soil that is exposed through digging (or in the sides of the soil cube). Their drawings will probably have layers like the drawing on page S27, but are not likely to be (or should not be) an exact replica. A copy of the soil profile on page S27 is provided as **Blackline Master Soil** 5.1 (Soil Profile). You may wish to make an overhead to use in your discussion with the student.

   Ensure that students pay attention to the way in which they record their observations. Encourage them to record every important detail in an organized way. If they do not have colored pencils or crayons with them, have them label their drawings to indicate the color of the soils and any variations that they see.

   Your students may not know that soil is layered in the ground (plant life at the top and materials being layered in a vertical sequence down). They may also be surprised by the depth of the soil. As they dig down into soil, it may seem that the soil layer is extremely deep. It is important that your students connect what they are doing in this investigation with the idea of Earth's crust. It is especially crucial that they realize that the soil layer, in terms of the thickness of the Earth's crust, is actually quite thin. For example, continental crust is 100,000 times the thickness of the average soil!

2. Give students time to write their predictions. This is another opportunity for students to apply what they already know to form a prediction. Again, it is important that students give clear reasons for their predictions to consider the predictions in the light of evidence. Encourage students to think quantitatively and descriptively. For example, "It will take 60 s for all of the water to move downward through one foot of soil. The reason behind my prediction is that it was very easy to dig the soil and the soil was loose, so there are lots of open spaces in the soil for water to move through."

   a) Student predictions may include "some of the water will go through the soil because soil has open spaces in it."

3. Sprinkling water from the watering can onto the ground adjacent to the soil pit helps students see how water moves through soil from the surface. Observe students as they revisit their predictions. Remind them to note differences between their predictions and their observations.

Teaching Tip
Pour water onto the grass in the area adjacent to the hole that you have dug or have a student do it for you. Immediately after you pour the water, take the shovel and make a new vertical slice through the wall of the hole about two inches in from the edge. This will enable students to observe and record the percolation of water down through the soil. If it has rained heavily recently and the soil is saturated with water, you may not be able to see the results.

    a) Students may note that the water does not flow through the soil in a straight line, that it moves through different layers at different rates, or that it has a harder time moving through the lower layers.

    b) Students may note that the water "disappears" into the ground and/or they may be able to observe a color change in the soil as the water moves through the soil.

    c) Answers will vary. Where student predictions do not work out, ask those students to consider the reasons they gave, and revise them based on the evidence of their observations.

4. Restoring the site (filling the hole and replacing the grass that was removed, if any) is very important for avoiding injury. Make sure that students assist with gathering all materials before heading back to the classroom.

Assessment Opportunity
Distribute an index card to each student and ask students to write on both sides, with these instructions:

Side 1
Based on your field study of a soil profile and how water moves through soil, list a big idea that you understand and word it as a summary of your understanding.

Side 2
Identify something about the soil profile or how water moves through soil that you do not yet fully understand and word it as a statement or question.

Review the cards for the information they provide about what your students have learned and the questions that they wish to explore further.

Teacher Commentary

NOTES

Investigating Soil

INVESTIGATING SOIL

Inquiry

Measurement in Scientific Inquiry

Scientists use measurement in their investigations. Accurate measurement, with suitable units, is important for both collecting and analyzing data. Data often consist of numbers.

Part 2: Designing and Conducting Your Own Investigation

1. Think about how water passes through the spaces between soil particles, as well as the different materials in soil and their sizes.
   Set up the testing equipment as shown. Use one set for each different test material.
   You will test:
   - plant material;
   - sand;
   - gravel;
   - clay.

   10 oz. clear plastic cup with hole in the bottom

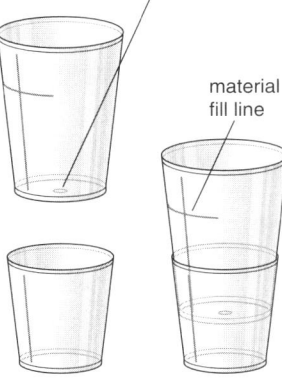

   material fill line

   6 oz. clear plastic cup to catch water

   Place the larger cup into the smaller cup.

2. In your group discuss how you could use the equipment and materials to develop a fair test.
   Your testing procedure should have:
   - a question to investigate;
   - a fair design;
   - a form of measurement that all can agree upon;
   - a clear method of recording observations and results;
   - a method that can be repeated to verify the results.

   a) Write down your procedure.

3. With the approval of your teacher, carry out your fair test.
   a) Record your data.

⚠ Have your plan approved by your teacher before you begin your investigation.

Design Investigations

Conduct Investigations

Collect & Review

S 28
Investigating Earth Systems

160 Investigating Earth Systems

Teacher Commentary

Part 2: Designing and Conducting Your Own Investigation

Having seen water moving through soil in the field, students observe in a lab setting how water moves through different materials that are found in soil. This is an opportunity to emphasize how observations made in the field can lead to laboratory experiments. This leads them into Part 3 of the investigation where students conduct percolation tests in the field.

1. Show students the materials for the investigation. If possible, give students an opportunity to look at the different materials through a binocular microscope so that they can observe differences in the material, such as grain size or amount of open spaces between grains of gravel, sand, and silt or between pieces of plant material. Grass (cut away from the topsoil) or peat moss, available at garden stores, are suitable materials for this test.

2. Here, students design and run a fair test within fairly simple parameters. By this point in the module, students should be able to figure out how to control the variables and make this a fair test. Encourage students to write down their procedure. Remember to collect student journals frequently! Use the checklists and rubrics in the **Appendices** for evaluating students work.

   From the materials provided, students will infer that they can test the rate of flow through the four substances. Samples of questions that students may generate include:
   - Which material will water run through the quickest?
   - How fast will water flow through plant matter, gravel, sand, and clay?
   - Will all of the water make it through all four materials?
   - Will more water pass through plant material than through gravel, sand, and clay?

   Stress that there is no "right" experiment. Emphasize that as long as they have a testable question, and a quantitative measure for answering their question (like rate of flow or volume), they can proceed with their investigation. Emphasize also that a fair design is one that keeps all factors the same—except the one they are testing. For example, if they are testing the four materials to see which one permits water to flow through fastest, they must do the procedure the same way with each of the four materials.

   a) Review students' experimental designs. Your review should focus on safety. Rather than pointing out flaws in their experimental designs, allow students to conduct their investigations and learn from the discussion that follows. Students should not begin unless they have written procedures (fair test procedures) and predictions.

Investigating Soil – Investigation 5

3. a) Water will move through the plant material almost instantly. It will move through gravel fairly rapidly, through sand less rapidly, and it will move very slowly through clay. Some students may be surprised that water moves differently through the different materials. They may be especially surprised at the variability in percolation rates.

Assessment Tool
Investigation Journal–Entry Evaluation Sheet
The will help students to learn the basic expectations for journal entries that feature the write-up of investigations. It provides criteria that students can use to ensure that their work meets the highest possible standards and expectations. Using and discussing the evaluation sheet will help your students to internalize the criteria for their performance.

When assessing student investigations, keep in mind that the **Investigate** section of an *IES* lesson corresponds to the explore phase of the learning cycle (engage, explore, apply, evaluate) in which students explore their conceptions of phenomena through hands-on activity. The way in which students design and conduct their tests here can be used as an assessment of their understanding of inquiry processes. This should alert you to any misunderstanding they might have and give you an opportunity to help them clarify their understanding.

Assessment Opportunity
Understanding Scientific Inquiry
Have your students fill in the blanks of the following analogy prompt as a means of checking their understanding of what it means to conduct a fair test in an investigation.

A fair test in a science investigation is like _____ because _____.

Teacher Commentary

NOTES

Investigating Soil

Investigation 5: Water and Other Chemicals in Soil

4. Share your findings with other groups in your class. Use everyone's results to answer the following questions:

   a) Which material lets water pass through it most quickly?

   b) How does the rate at which water passes through a material relate to the size of the particles in the material?

Part 3: Conducting a Percolation Test

1. Scientists use the term percolation to describe water passing through materials.

   Follow these steps to conduct a percolation test.

   1 Obtain a large soup can that has both ends removed.

   Mark a spot halfway up on the side of the can.

   Mark the halfway spot on the empty soup can.

   2 Select the site you want to test.

   Make sure the soil there is not too dry or hard. If necessary, water the ground first.

Clean up any spills immediately.

Inquiry
Following Standard Procedures

You are going to run a percolation test on soil. This test is similar to the one professionals use in the field to find the rate at which water moves through soil. You want to be able to compare your results with others. Therefore, it is important that all tests be done using exactly the same procedure.

Investigating Earth Systems

S 29

164 Investigating Earth Systems

Teacher Commentary

4. You will need to organize the sharing of results. One way is to ask a spokesperson from each group to come to the front of the class and give a presentation. This will encourage groups to display their results in a clear way for others. Alternatively, you could have groups show results to each other in a series of pairs. Since everyone has been experimenting with the same materials and equipment, you can expect similar results from each group. Differences are likely to be in the design of the tests. Where differences appear, have students discuss possible reasons for them. The result of the sharing and discussion should be some clear answers to the questions listed here.

   a) Consensus should be easily reached on this question – water flows fastest through plant matter. This assumes that your students use grass or peat moss and do not pack the material down into the cup.

   b) In most cases, there is an inverse relationship between size and rate of flow (i.e., small particles let things through slowly, big particles let solutions through quickly. In the case of plant materials, it may depend on the type of plant materials you use.

> **Teaching Tip**
> Have students work ahead by beginning to research local soil properties. Students should look for any information on local soil types. The Internet, books, and magazines can all be used as sources. Local gardeners are excellent sources of information as well. Provide students with addresses that they can write to for free garden supply catalogs. You might find these in gardening magazines. Inform students that they will need these catalogs in Investigation 7: Using Soil Data to Plan a Garden.

Part 3: Conducting a Percolation Test

1. It is important that all students fully understand what this test is designed to do. If necessary, help students by sketching on a blackboard or chart paper what should happen. Have students read the four steps in the procedure. Answer any questions that they have about safety protocols or percolation test procedures before they go outdoors.

   Emphasize to students that they must all follow the same procedure. Ask them why this is necessary. Work with your students to prepare a data chart before they conduct their perc tests. It is important that the chart is big enough to include data from other groups. You might find it helpful, if your students have trouble organizing data charts, to make up a whole class chart that the students can adapt for their own uses.

Teaching Tip

As before, you will need to find places within or near your school grounds where these tests can be conducted. If possible, try to pick sites that are similar to those used in Investigation 4 for soil core sampling. This will allow for additional comparisons to be made.

Percolation can happen quickly or slowly, depending on the soil type and recent weather conditions. Try the perc test yourself ahead of time, so that you can make adjustments to the procedure to suit your soil conditions.

Safety

The four steps in the procedure are illustrated and explained in the Student Book. You might feel that it would be best to hammer the soup cans into the ground yourself. Use your judgment about this. When hammering the can into the ground, the hammer should strike the wood block head on and not at an angle. Soup cans are fairly fragile. Use sturdy, steel cans. In very hard ground they may buckle or bend. Large stones or tree roots can also be a problem. It will be worth having some spare cans available. You will need to find good places to drive the cans into the ground. A pointed stick pushed into the ground will tell you where good spots are for the test.

Teacher Commentary

NOTES

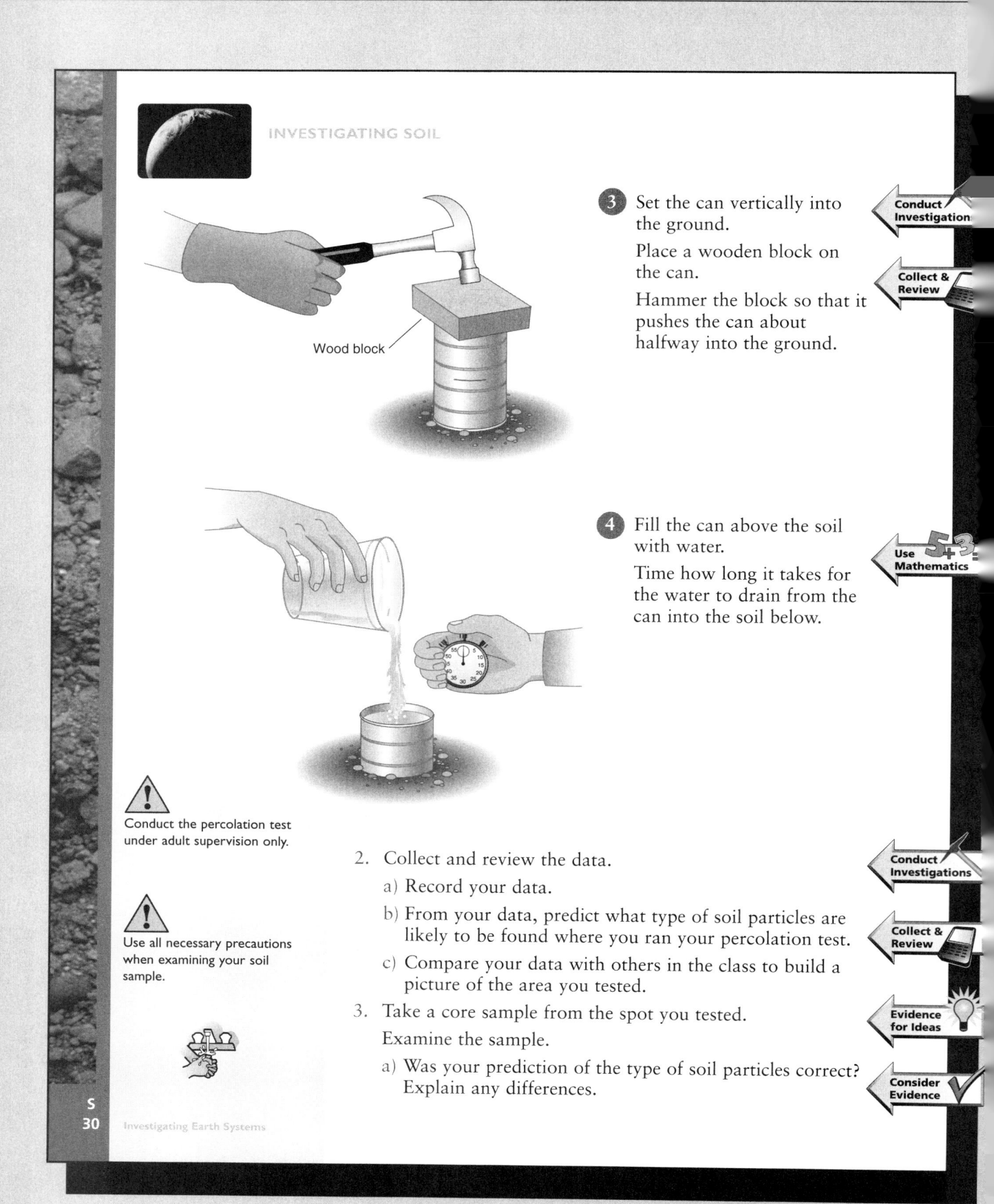

Investigating Soil

INVESTIGATING SOIL

Wood block

3. Set the can vertically into the ground.

   Place a wooden block on the can.

   Hammer the block so that it pushes the can about halfway into the ground.

4. Fill the can above the soil with water.

   Time how long it takes for the water to drain from the can into the soil below.

⚠ Conduct the percolation test under adult supervision only.

⚠ Use all necessary precautions when examining your soil sample.

2. Collect and review the data.

   a) Record your data.

   b) From your data, predict what type of soil particles are likely to be found where you ran your percolation test.

   c) Compare your data with others in the class to build a picture of the area you tested.

3. Take a core sample from the spot you tested. Examine the sample.

   a) Was your prediction of the type of soil particles correct? Explain any differences.

S 30

Investigating Earth Systems

Teacher Commentary

2. Observe students as they conduct the percolation tests.

   a) – c) Encourage students to write down their data and their predictions. Have students share data by giving them a list for their names and their percolation times. Ensure that it gets passed around to all groups. Check to make sure students write down their predictions before taking core samples.

3. After your students have finished their percolation tests, they are asked to take a core sample and to examine the sample. The sample should be taken in the center of the soup can. See Investigation 4 for instructions on taking core samples.

   Ask students to think about any possible relationships between the two.
   - Does the structure of the core sample have anything to do with the percolation time? How could that be so?
   - What other factors might affect the percolation time?
   - Can your students think of any way in which a percolation time could be sped up or slowed down?

   a) Silt, sand, and clay exist in different proportions in different soils. This gives soils unique characteristics and texture. For example, a soil can have a sticky character (when wet) or soft and velvety with grittiness (when dry) if it contains mostly clay particles (the smallest-sized particles of the three). It can be velvety smooth with a hint of grit if it has a high silt content (medium-sized), and gritty if it contains mostly sand (the largest-sized particles of the three). Students need to look for these characteristics, which may affect the results of percolation tests.

Teaching Tip
Keep some of the core samples for Part 4 of this investigation. Otherwise, you will need to collect new soil samples for the pH and nutrient-testing exercise.

Assessment Opportunity
Ask students to produce a concept map or flow chart to show the relationships between the types or size of materials in soil and the rate at which water flows through these materials. If you wish to provide prompts to help them to develop the map, here are a few suggestions:

Title: Relationship between Soil Material and Percolation Rate.

Materials: plant material, gravel, sand, clay.

Connectors: water flows fastest through..., water flows slowest through..., because... (and so on).

Investigating Soil

Investigation 5: Water and Other Chemicals in Soil

Part 4: Testing for Other Chemicals

1. Obtain a soil-testing kit. Examine the kit carefully. Read to find out:
   - what different chemicals the kit measures;
   - why these chemicals are important in the soil;
   - what testing procedures have to be followed;
   - any safety precautions that should be observed.

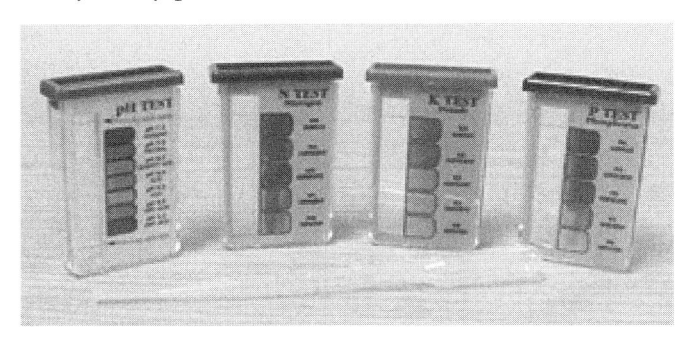

   > Materials in soil-test kits can be harmful if swallowed. Be sure you understand all safety precautions and handle the chemicals responsibly.

2. Run the tests as directed on the kit. Remember, it is always a good idea to repeat your tests several times to get reliable results.

   a) Record your results in a form that others can see and understand.

3. In your group analyze your results.

   a) What chemicals are at a high level in your sample? What chemicals are at a low level?

   b) Combine these results with what you found out earlier about water, solid particles, plant life, and other aspects of your soil samples. This information will be very useful in (the final) Investigation 7.

   c) Pool all the data collected by the class. Add all new items to your soil maps.

Teacher Commentary

Part 4: Testing for Other Chemicals

In this investigation, students will extend their inquiry experiences by conducting tests and by using a specialist's tool (soil-testing kit) to measure for pH and soil chemicals. It is not likely that many students will have done this previously. All the soil tests use color-changing indicators. Help students to see that indicators are an important tool in science.

> **Teaching Tip**
>
> Soil-test kits are commonly available at garden and hardware stores. The instructions and information they contain on their packaging is usually well written and easy to understand. Before you purchase soil-test kits, get a sample one and review the background information it provides. Some may be more sophisticated than others. Pick one that you feel your students are most likely to understand.

1. Most kits come with some sort of reading which includes this information. Having students find the information from the kits themselves is an important learning experience. It also helps them see how scientific ideas and concepts are linked to the real world. An important part of science as inquiry is gathering information from reliable sources. Interpreting scientific information from these soil kits is a good example of this.

    Test kits vary slightly in their instructions; however, nutrient tests differ from pH in that nitrogen, potassium and phosphorus are bound tightly to soil particles. Virtually all test kits require you to make a soil solution. Most recommend that the soil solution sit out overnight for better extraction of nutrients. To make a soil solution, add about 1 part soil for every 5 parts distilled water. This mixture should be shaken vigorously. Overnight, particles will settle. The extracted nutrients sit on the top layer of water. This is the material that you will use for your test.

    The three chemicals that most soil test kits indicate are: nitrogen, phosphorus, and potassium.
    - Plants can absorb nitrogen when it is in the form of nitrate (NO_3^-) or ammonium (NH_4^+).
    - Plants also need phosphorus (phosphates) and potassium (potash) to ensure healthy growth.
    - Different plants need different proportions of these three nutrients to thrive.

2. Run the pH test first. It is the easiest and quickest test to perform. Check to ensure that students are following safe laboratory procedures.

    a) Encourage students to record their results in an easy-to-understand way.

Investigating Soil – Investigation 5

> **Making Connections ...with Chemistry**
>
> This is an opportunity to introduce and discuss pH. You will have to decide how deeply to discuss acids and bases and the pH scale. Some students who have pools or spas may already be familiar with pH. They may, however, have little idea of the science involved beyond keeping the water safe for use.

3. Students may need your help in pulling together all the evidence they have collected. Some groups may have been more successful than others in interpreting the information from the soil-test kits. Recognizing differences in data is more important than a deep understanding of the chemistry. Students should be able to grasp the fact that different plants need different chemical conditions to thrive, and that the soil tests indicate the levels of the three key chemicals involved.

   a) Results will vary. You may want students to write their results on the board or on the overhead projector. Create a chart so that each group can easily record its data. Encourage students to copy this chart.

   Although your students have explored soil in many ways, this is the first time they will have focused on chemicals. In addition to water, they will be investigating other chemicals, specifically nitrogen, potassium and phosphorus. Students might have little or no understanding of the role these chemicals play in plant growth. They may appreciate the importance of fertilizer; however, they may not think of fertilizer as a natural chemical. Furthermore, they may not appreciate that a fertilizer may be good for one type of plant but bad for another. Ensure that students realize that water is a chemical substance made of hydrogen and oxygen. Remember, some students may think water is somehow chemically neutral.

Teacher Commentary

NOTES

Investigating Soil

INVESTIGATING SOIL

As You Read...
Think about:

1. Why is it important to know the drainage rate of soil?
2. Why are water and other chemicals in soil important?
3. What is pH level and why is it important to plants?
4. What chemicals are important to plant growth?

Digging Deeper

The Importance of Water and Other Chemicals in Soil

Knowing how water passes through soil is very important. For example, engineers need to know how quickly water will drain away from buildings or bridges. Drainage rates are also important to farmers, landscapers, gardeners, environmental scientists, and other professionals.

Water is one chemical found in soil. There are many others. Some of the chemicals in soil are very important for plant growth. Farmers and gardeners often test their soil for these chemicals. They use soil-testing kits. These kits can be bought at gardening or hardware stores.

Evidence for Ideas

Teacher Commentary

Digging Deeper

Instruct students to read the **Digging Deeper** section.
The first paragraph provides relevance to the percolation tests that students completed in Parts 1 and 3 of the investigation. The remainder of the text on pages S32 and S33 focuses on helping students to understand soil chemistry.

> **About the Photo**
>
> The photograph on S32 shows what happens when rainfall is so heavy that it exceeds the capacity for the soil to absorb it. (This is especially problematic in developed areas where concrete and asphalt prevents water from entering the ground.)

As You Read...

Tell students to complete questions 1–4 in their journals.

1. One reason that it is important to know the drainage rate of soil is that plants need a continuous supply of water, which often must suffice for long periods after a rain. The slower the drainage rate, the longer soil water will be available for plants. Other reasons have to do with knowing how fast water will drain away from buildings and bridges.

2. Water and other chemicals are important in soil because they control the pH and nutrients available to plants.

3. The pH level refers to how acidic or basic a solution is. The pH of the soil controls how well plants use the food (nutrients) available in the soil.

4. The most important plant nutrients are nitrogen, phosphorus, and potassium.

Investigation 5: Water and Other Chemicals in Soil

Soil-testing kits can test to see how acidic or basic a soil is. (Vinegar is acidic, and many soaps are basic.) The measure for how much acid or base there is in soil is called the pH level. A pH of 7 means that there is no more acid than base present. The solution is neutral. Lower numbers indicate acids. Higher numbers indicate the presence of bases. The pH controls how well plants use the food (nutrients) available in the soil. Different plants prefer different pH levels. The testing kit you use may tell you which plants do better at a given pH. The pH level in soil can be changed by adding chemicals.

Three chemicals are very important for plants. They are:
- nitrogen that affects the growth of leaves;
- phosphorus that helps roots grow strong;
- potassium that helps flowers and fruits grow.

Soil-testing kits can tell you how much of each chemical there is in a soil sample. Different plants need different amounts of these chemicals. Lawn grasses need a high level of nitrogen. Fruits need less nitrogen but higher levels of phosphorus and potassium. Root vegetables need lower levels of nitrogen and potassium and much higher levels of phosphorus.

Teacher Commentary

Assessment Opportunity
You may wish to select questions from the **As You Read** section to use as quizzes, rephrasing the questions into multiple choice or "true/false" formats. This provides assessment information about student understanding and serves as a motivational tool to ensure that students complete the reading assignment and comprehend the main ideas.

Teaching Tip
For additional teaching suggestions on how to present the content of this **Digging Deeper** reading section to your students, please refer to the *IES* web site.

Investigating Soil

INVESTIGATING SOIL

Review and Reflect

Review

1. Look again at the key question. Has the investigation helped you answer the question? Based on your investigation, what chemicals are in your soil sample?
2. Suppose you were to conduct percolation tests on soil that contained a lot of clay and soil that contained a lot of gravel. Which would you predict would take less time?
3. Based on the result of your investigation and your reading, explain why it is important to know what chemicals are in soil.

Reflect

4. Where do you think the chemicals in soil come from?
5. How could you use what you know about soil in a practical way?
6. Based on what you now know, what further questions about chemicals in soil would be useful to investigate?

Thinking about the Earth System

7. How do water and other chemicals in soil connect to the geosphere, hydrosphere, atmosphere, and biosphere? Include any new connections on your *Earth System Connection* sheet for soil.

Thinking about Scientific Inquiry

8. Describe how you used evidence to develop ideas in this investigation.
9. Why was it important for everyone to follow the same procedure exactly when doing the percolation test?
10. Explain where you used measurement and why it was important in this investigation.

Teacher Commentary

Review and Reflect

Review
It is important that students carefully review what they have done in this investigation, especially those events that have extended their understanding of soil. In particular, students need to relate the chemical testing to their earlier investigations. Spend time helping students review what new things they have learned about soil.

1. Answers will vary.

2. The soil with a lot of gravel (because it is likely to be more porous and more permeable).

3. Chemicals in soil can act as plant nutrients, or, especially in high concentrations, plant poisons.

Reflect
Question 6 provides another opportunity for students to realize that investigations tend to generate new questions for inquiry. Again, they may need your help in seeing that this is part of the nature of scientific inquiry.

4. The chemicals in soil come from rocks, minerals, precipitation, living things, and the atmosphere.

5. Answers will vary. Planting a garden or building a house are two examples.

6. Answers will vary. Examples include: Which soil chemicals are best for growing vegetables? What happens to plants when the pH of the soil is raised or lowered?

Thinking about the Earth System
Students should now have a better understanding of all four of these Earth Systems. Remember, it is the interconnections between these systems that students need to appreciate. They should be able to see more clearly that soil in the geosphere is very closely linked to water in the hydrosphere, and to living things in the biosphere, as well as to the air in the atmosphere.

7. Water can move through the geosphere at different speeds depending on the type of soil present. Plants depend on various types of soil nutrients for their survival in the biosphere. Chemicals in the atmosphere influence soil pH. Water dissolves chemicals; acidic water dissolves more chemicals than neutral water.

Thinking about Scientific Inquiry
Discuss **Review and Reflect** question 9 in class. This is an important part of inquiry. Have students cite examples of how everyone in the class followed the same procedure. Let students explain why this is important.

Students may again need your help to see that they have been investigating soil through a set of clear scientific processes that these are generic to many investigations. Students' attention needs to be drawn to the processes they have used and how and when they used them. Help students to appreciate the importance of inquiry processes by reviewing the **Blackline Master** of **Inquiry Processes**. Take time to discuss the inquiry procedures that your students used in this investigation. The investigation has been a combination of field and laboratory testing. Help students to see the connections.

8. Predictions were made regarding the flow of water through plant material, sand, gravel, and clay. Evidence was gathered to show the relative speeds of flow through these materials.

9. In order to compare percolation rates, all other variables (amount of water used, type of can used, and so on) must be the same. Otherwise, the percolation rate will be different because of those factors, not the factor we care about—soil particle size.

10. Measuring time during the percolation test; used to infer the size of soil particles in the ground. Measuring levels of chemicals and pH; plants have different needs regarding pH and nutrients. The amount of potassium, phosphorus, and nitrogen in the soil, along with the pH of the soil, determine which plants can live there.

> **Assessment Tool**
>
> **Review and Reflect Journal–Entry Evaluation Sheet**
> Depending upon whether you have students complete the work individually or within a group, the **Review and Reflect** portion of each investigation can be used to provide information about individual or collective understandings about the concepts and **Inquiry Processes** explored in the investigation. Whatever choice you make, this evaluation sheet provides you with a few general criteria for assessing content and thoroughness of student work. Adapt and modify the sheet to meet your needs. Consider involving students in selecting and modifying the criteria for evaluating their end of investigation reflections.

Teacher Commentary

NOTES

Teacher Review

Use this section to reflect on and review the investigation. Keep in mind that your notes here are likely to be especially helpful when you teach this investigation again. Questions listed here are examples only.

Student Achievement

What evidence do you have that all students have met the science content objectives?

_____

Are there any students who need more help in reaching these objectives? If so, how can you provide this? _____

_____

What evidence do you have that all students have demonstrated their understanding of the inquiry processes? _____

_____

Which of these inquiry objectives do your students need to improve upon in future investigations? _____

_____

What evidence do the journal entries contain about what your students learned from this investigation? _____

_____

Planning

How well did this investigation fit into your class time? _____

_____

What changes can you make to improve your planning next time? _____

_____

Guiding and Facilitating Learning

How well did you focus and support inquiry while interacting with students?

_____

What changes can you make to improve classroom management for the next investigation or the next time you teach this investigation? _____

_____

Teacher Commentary

How successful were you in encouraging all students to participate fully in science learning? _____

_____

How did you encourage and model the skills values, and attitudes of scientific inquiry? _____

_____

How did you nurture collaboration among students? _____

_____

Materials and Resources

What challenges did you encounter obtaining or using materials and/or resources needed for the activity? _____

_____

What changes can you make to better obtain and better manage materials and resources next time? _____

_____

Student Evaluation

Describe how you evaluated student progress. What worked well? What needs to be improved? _____

_____

How will you adapt your evaluation methods for next time? _____

_____

Describe how you guided students in self-assessment. _____

_____

Self Evaluation

How would you rate your teaching of this investigation? _____

_____

What advice would you give to a colleague who is planning to teach this investigation? _____

_____

NOTES

INVESTIGATION 6: SOIL EROSION

Background Information

Erosion is the wearing away of soil or rock. Soils, being unconsolidated materials consisting of particles that are not cemented or bonded to one another, are very susceptible to erosion. Soil is eroded by two agents: running water, and wind. The processes of soil erosion by water and by wind are similar but are different in important ways.

When a fluid like air or water passes over a surface covered by rock and mineral particles, the fluid exerts forces on the particles. The basic nature of these forces is much the same as the forces involved in settling of particles through a fluid, discussed in the **Background Information** for Investigation 2. These forces tend to mobilize the particles and carry them downstream or downwind. The forces even have a component of lift, for the same reasons that an airplane wing experiences lift! The stronger the flow of fluid, the greater the forces on the particles. When the flow reaches a certain speed, called the threshold speed or the critical speed, some of the particles are moved. Movement of sediment in this way is the rule in rivers and streams, especially during floods, but it is the exception on soil surfaces.

When rain falls on the land surface, some of the water infiltrates into the soil, but if the rain is heavy, some or even most runs off across the land surface as overland flow. Overland flow tends to be slow, and is not effective in eroding and transporting soil particles, but as it proceeds downslope it tends to become concentrated in low areas and become channelized. Channelized flow is deeper and faster than overland flow, and it is more likely to cause erosion. In areas with scant vegetation and very erodible soil, the entire land surface becomes channelized, with channels as small as a decimeter across. Such areas are called badlands. Many badlands (which, despite the name, often have considerable natural beauty) were experiencing natural soil erosion long before human habitation. In areas where all of the soil is covered by vegetation, on the other hand, the erodibility of the soil is very low, and channelization is not effective. If such areas are put under cultivation, however, channelization can develop during times when the ground is devoid of either crops or natural vegetation. Once a channel is initiated, it can grow rapidly in depth, width, and upslope position. Such channelization is the cause of most soil erosion by water. The products of soil erosion by running water are carried away in suspension during times of strong flow, and eventually they are delivered to large rivers downstream. When you see a stream or river running brown during a flood, you can be sure that soil erosion is active somewhere upstream.

Strong winds exert forces that are sufficient to erode dry soil particles resting on bare soil surfaces. Natural wind erosion in dry desert areas is limited by availability of fine material on the surface. It is common only after infrequent heavy rains spread a fresh layer of fine sediment on the land surface. As the fine material is eroded away by the wind, the land surface becomes armored by the coarser material as it becomes more and more concentrated at the surface protecting the remaining fine material beneath from the effects of the wind. In agricultural soils, however, if the soil becomes dry for long periods of time and unusually strong winds blow them as well, enormous quantities of fine soil material can be suspended by the wind and carried for hundreds or even thousands of miles before settling out or being washed out by rains. There is usually too little coarse material in such soils to produce an effective layer of armor. Soil erosion by the wind was a serious problem in the central and southern Great Plains in the United States in the "Dust Bowl" times of the 1930s. The areas were put into agriculture during earlier times of abundant rainfall, and when a long period of drought followed, the soil, now stripped of its natural cover of prairie grasses, was especially vulnerable to wind erosion.

Soil erosion by water can be minimized by such agricultural practices as planting rows along topographic contours (called contour farming), to minimize downhill flow of runoff, and keeping bare soil covered by use of cover crops. Many farmers are now using "plowless" planting, which minimizes exposure of bare soil to erosion. A cover of vegetation is also very effective in preventing wind erosion. The best way to prevent wind erosion, however, is to not put lands in vulnerable semiarid regions under cultivation in the first place, if it can be avoided.

The *Investigating Earth Systems* www.agiweb.org/ies web site also contains a variety of links to web sites that will help you deepen your understanding of content and prepare you to teach this investigation.

Teacher Commentary

Investigation Overview

Having studied how soil forms and how water passes through soil, students explore how soil is worn away. They use a stream table to investigate soil erosion. They begin by exploring the model, examining the relationship between the flow of water and erosion rates, and between particle size and erosion. Students use their initial explorations to develop their own questions and investigations. In Part 2, students then study soil erosion by wind. Again, students begin by exploring the model, examining the relationships between wind speed and erosion rate, and comparing and contrasting water and wind erosion. Students use their understanding to form and test their own questions about soil erosion. Text provides further explanation about the nature of soil erosion and steps that people can take to reduce it.

Goals and Objectives

As a result of this investigation, students will develop a better understanding about the nature of soil erosion by water and wind, and improve their ability to use models to design and conduct scientific inquiry.

Science Content Objectives

Students will collect evidence that:
1. Wind and water that move across a soil surface apply forces to soil particles.
2. The stronger the water current or wind, the more particles that are put into motion.
3. Proper planning can reduce soil erosion.
4. Soil erosion is a serious problem in many areas.
5. Predictions can be derived from models and these predictions can be tested.

Inquiry Process Skills

Students will:
1. Identify questions that can be answered through scientific investigations.
2. Design and conduct a scientific investigation.
3. Use appropriate tools and techniques to gather, analyze and interpret data.
4. Develop descriptions, explanations, predictions, and models using evidence.

Connections to Standards and Benchmarks

In this investigation, students will investigate how water and wind erode a natural resource (soil) that takes thousands of years to form. These observations will start them on the road to understanding the National Science Education Standards and the American Association for the Advancement of Science Benchmarks shown below.

NSES Link

- Human activities can induce hazards through resource acquisition, urban growth, land-use decisions, and waste disposal. Such activities can accelerate many natural changes.

AAAS Link
- Models are often used to think about processes that happen too slowly, too quickly, or on too small a scale to observe directly, or that are too vast to be changed deliberately, or that are potentially dangerous.

Teacher Commentary

Preparation and Materials Needed

Preparation

Part 1: Soil Erosion by Running Water
This investigation requires a stream table or large plastic tray to model stream erosion by running water. The experimental setup is shown on page S36 of the Student Book, but adapt this to suit your particular needs and the materials you have available. Ensure that you have adequate water sources for this experiment. Running a hose from a sink to the container filled with soil is the ideal arrangement, but pouring water from a container onto the sponge will also work. Your students will need to collect the water that exits the bottom of the container. If you do not have large containers to catch the water, you can place the "stream table" container inside a larger container and use the larger container to collect the water.

In Step 5 of the investigation, students develop their own questions to investigate using the stream table, water, and soil. They are asked to think about methods for reducing soil erosion. You might consider having some extra materials available that they can use, such as peat moss (to simulate vegetative cover).

Part 2: Soil Erosion by Wind
This activity is best performed outdoors because airborne soil particles will create a mess in the classroom and create dust that you do not want students to inhale. The challenge will be to find a location outside yet near the classroom so that an extension cord can be run to the fan. Common sense dictates that you should not work with electric fans and extension cords outdoors in inclement weather (e.g., rain, sleet, snow, and/or excessive wind) or when the ground is wet.

You will need a supply of dry soil because wet soil is far less likely to be eroded by wind than is dry soil. Break up the soil before spreading it out on the plastic drop cloth. Students should stay upwind or to the side of the soil while the fan is blowing, and should have dust masks and safety goggles to minimize the likelihood of inhaling dust from soil. Ideally you should test the experimental setup prior to performing the experiment with students. You might wish to bring a meter stick with you for recording measurements of the transport of soil.

Given the demands on school schedules and the vagaries of doing outdoor field work (especially when the weather does not cooperate), you can consider the following alternative arrangements:

1. Conduct Steps 1 – 4 of the investigation as a demonstration (either with your students or by running the experiment yourself and videotaping the results and then showing the video in class). You could then assign students to develop investigations (Step 5) and assign Step 5 as an assignment to be completed at home or for extra credit.

2. You can also assign the entire activity (Steps 1 – 5) as homework to be completed in the yard under adult supervision.

Materials

- stream table or large plastic tray with end cut open
- soil sample to fill the tray
- tube or hose connected to a water faucet
- large garbage pail
- small piece of wood or other item to prop up tray
- large sponge
- metric ruler
- large sample of dry soil, about 5 kg (10 lb.)
- electric window fan, or a fan on a stand
- plastic drop cloth
- weights to keep the drop cloth in place
- dust masks
- safety goggles

Teacher Commentary

NOTES

Investigating Soil

Investigation 6: Soil Erosion

Investigation 6:
Soil Erosion

Key Question
Before you begin, first think about this key question.

How is soil worn away?

Think about what you already know about how soil is formed. What factors could act on soil to make it wear away? Share your thinking with others in your group and with your class.

Materials Needed:

For this investigation your group will need:

- stream table or large plastic tray with end cut open
- soil sample to fill the tray
- tube or hose connected to a water faucet
- large garbage pail
- small piece of wood or other item to prop up tray
- large sponge
- large sample of dry soil, about 5 kg (10 lb.)
- electric window fan, or a fan on a stand
- plastic drop cloth
- dust masks

Investigate

Part 1: Soil Erosion by Running Water

1. Set the stream table on a table top with the open end hanging over the edge.
   Place the pail under the open end of the tray at the edge of the table. Fill the tray with the soil sample to a depth of about 2 to 3 cm (1").

Investigating Earth Systems

S 35

192 Investigating Earth Systems

Teacher Commentary

Key Question

Instruct students to respond to the **Key Question** in their journals. Allow a few minutes of writing time. Have them share their ideas with a neighbor, then with the rest of their group, and finally, the entire class. Make a list of ideas on the board. Accept student ideas uncritically, even if they appear undeveloped (or are simply not correct). Neither should wrong ideas be praised. The point of this exercise, as it is with all the **Key Questions**, is to provoke thought and prepare for the investigation.

Instruct students to make a "master list" in their journal of ideas about how soil is worn away.

Student Conceptions

Students are likely to identify rain, wind, and rivers (running water) as the main agents that wear soil away. Other conceptions include walking on soil, or landslides.

Answer for the Teacher Only

Wind and water are the two main agents that erode soil. Both agents act more vigorously on soil when soil is exposed (stripped of vegetation or where plants do not grow).

> ### Assessment Tool
> Key–Question Evaluation Sheet
> The **Key–Question Evaluation Sheet** will help students to understand and internalize basic expectations for the warm-up activity.

Investigate
Teaching Suggestions and Sample Answers

Part 1: Soil Erosion by Running Water

1. Have students read all steps of the investigation, then hold a class discussion. In the discussion, review the experimental procedure with students to check that they understand what they are supposed to do and why they are doing it.

   If the materials and models that you will use differ from what is shown on page S36, point out the differences and the important steps to using the models. (The diagram of the setup is provided as **Blackline Master** *Soil* 6.1 (Stream Table Setup) if you wish to make an overhead for discussion purposes.) One is that water must be poured or sprayed onto the sponge (Step 2), which breaks the force of the water jet and makes the model a more realistic simulation of erosion by running water. The second point is that students must be careful to collect water at the lower end of the stream table because water that spills onto the floor creates a safety hazard. One student in each group should assume the primary responsibility for safety. The third point is that students will need to make careful observations of the model and how it changes (Steps 2 – 4). Encourage them to think about how to do this by pointing out the questions in Step 4 of the investigation. The following questions, discussed prior to the investigation, might also be useful:

- What do you need to observe? (*How soil particles are moved by the water, how the movement of coarser particles differ from the movement of smaller particles, whether erosion is concentrated in one place.*)
- How can you make a record of what happens to the soil as a result of running water? (*Observe the model carefully and write down notes about what happens, focus on a particular piece of soil and watch what happens to it, etc.*).
- How can you document how the soil in the model changes over time? (*Draw before and after pictures of the model – before water runs through it and after water has stopped running through it.*)

Finally, make sure that students understand that their observations in Steps 2 – 4 are meant to generate new questions for inquiry in Step 5, where students design and conduct their own investigation into soil erosion.

Teaching Tip
An alternative for collecting water from the stream tables is to tape trash bags (with heavy-duty duct tape) to tables and to place the bags inside of large cardboard boxes (which provide support). These can catch runoff from the stream tables.

Once you are comfortable that students understand the investigation, allow them to proceed.

Teacher Commentary

NOTES

INVESTIGATING SOIL

Inquiry
Using Models in Scientific Inquiry

Scientists often use models as a way of studying and explaining things that cannot be observed directly. Modeling can also simulate processes that take a very long time in the real world. You will use a stream table to model soil erosion.

Explorations

Not all experiments are as well planned as the ones you have been conducting in the previous investigations. Sometimes, scientists do experiments just to watch and think. They then develop ideas for further experiments. This is a kind of exploration. Exploration is an important part of scientific inquiry. The experiments you are doing in this investigation are explorations.

⚠️ Arrange the stream table so that water flow can be stopped easily in case it begins to overflow. Clean spills immediately.

Level the surface of the soil and pack it down gently.

Raise the closed end of the tray by putting the small piece of wood under it.

Put the sponge on the soil surface at the upper end of the tray.

2. The idea is to explore what flowing water does to the soil. To do this, hold the end of the hose a short distance above the sponge and very gradually increase the flow of water. The purpose of the sponge is to break the force of the water jet that comes from the hose.

   Watch the soil surface in the tray carefully until some of the soil particles are moved by the flow of water.

   a) Take written notes, in as much detail as you can, about how the soil particles are moved by the water.

3. Increase the water flow slightly, and make more observations.

   Continue increasing the flow and making observations.

   a) Record all your observations.

4. Use your observations to answer the following questions:

   a) How did the movement of the coarser soil particles differ from the movement of the finer soil particles?

   b) Is the soil erosion concentrated in one place, or does it affect the whole width of the tray?

   c) Are some of the soil particles too large or too heavy to be moved by even a strong water flow?

Teacher Commentary

2. Observe students throughout the investigation, checking for safety and to ensure that they are recording their observations.

   a) In addition to written notes, students should make drawings of this step.

> **Teaching Tip**
>
> Instruct students to let water flow through the hose very slowly. There is no need for great water pressure in this activity. If spills occur, wipe them up quickly.

3. Encourage students to note in their journals whenever there is a change in procedure, in this case a change in the flow rate of water on the stream table.

   a) Students should observe a greater amount of soil erosion with increased flow rate of water. They should also note that larger particles that may not have moved when the rate of flow was lower now begin to be dislodged and transported down the stream table.

4. Instruct students to discuss these questions within their group before answering them.

   a) Students should observe that coarse particles do not move as far nor as fast as fine particles. Also, coarser particles are more likely to bounce, roll, stop, then move again, whereas smaller particles are more likely to be carried in or on the water (suspended).

   b) Observations will vary depending upon how students conduct the investigation. The location where water hits the soil is typically affected the most, and that the effect of water decreases with distance from this spot. Over time, most of the bed of soil will be affected and disturbed in some way by the flow of water.

   c) Students may answer yes or no. Prompt them to explain their answers. Very strong water flow can move even extremely heavy particles; however, student observations may vary depending on stream table setup.

> **Teaching Tip**
>
> In Step 5 of the investigation, students are expected to draw upon what they observed and learned about soil erosion to formulate new questions for inquiry. They may need your help in doing this. After students complete Step 4, hold a class discussion about the results of their work. Point out the notes about inquiry in the margin of page S36 titled **Explorations**. The text explains how scientists use experiments to develop ideas for further testing.

Assessment Tool

Investigation Journal–Entry Evaluation Sheet

This sheet will help students to learn the basic expectations for journal entries that feature the write-up of investigations. It provides a variety of criteria that both you and students can use to ensure that student work meets the highest possible standards and expectations. Adapt this sheet so that it is appropriate for your classroom. You may also wish to make modifications to develop a sheet specific to this particular investigation. Be sure to discuss the intent of the sheet and the criteria before applying it for grading purposes.

Teacher Commentary

NOTES

Investigation 6: Soil Erosion

5. In your small groups discuss further investigations you would like to do based on your exploration. For example, you may study ways to prevent or lessen the erosion of the soil by the flowing water.

   Share your ideas with the class.

   Design an investigation to test if your predictions were correct.

   a) List the materials you will need, the steps of the investigation, and your prediction. Don't forget to give a reason for your prediction.

   b) With the approval of your teacher carry out your investigation. Record your findings. Were your predictions correct?

Part 2: Soil Erosion by Wind

1. As a class, go outside to an open area on a nice day with little wind.

   Spread the drop cloth out on the open area. Weigh the corners down with bricks or heavy books.

   About one-third of the way in from one edge of the drop cloth, spread a patch of the soil sample about 1 cm thick. The soil should cover an area about 60 cm (2') on a side.

   Place the fan on the drop cloth about 1 m (3') away from the edge of the patch of soil.

   Aim the fan across the soil sample and toward the center of the drop cloth. Direct the fan slightly downward toward the soil.

Teacher Commentary

5. Keep stream tables out for student use. Encourage students to discuss their ideas and to record their questions and ideas. Consider holding a class discussion about students' questions for inquiry. Before students begin to conduct new investigations, check to see that they have a testable question.

> **Teaching Tip**
> **Making Predictions**
> This is another opportunity for students to apply what they already know to form a prediction. Again, it is important that students give clear reasons for their predictions to consider the predictions in the light of evidence. Where student predictions do not work out, ask those students to consider the reasons they gave, and revise them based on the evidence of their observations.

   a) Insist that students have a mechanism for evaluating the results of their experiment. How will they measure soil erosion? Examples of predictions that students might make include:
   - Compacted soil will erode more slowly than loose soil because it will be harder for the water to pick up particles of soil.
   - Covering soil with plants (e.g., peat moss) will reduce erosion because the plants will hold down the soil or protect it from the water.
   - Less soil will erode on a low slope than on a steep slope because water will move more slowly on a low slope.

   b) Observe students as they investigate. Make sure that students record and explain their findings.

Part 2: Soil Erosion by Wind

As with Part 1, have students read all steps of the investigation. After students have read, check for understanding of the intent and purpose of the investigation. The best way to do this is to take students through the questions 2. a), 3. a), and 4. a) – d), asking students about what they will be investigating, what they expect to happen, what they will need to record, how they will gather evidence about what happens, and so on.

Review protocols for safety and behavior. Students should wear safety goggles and dust masks during this investigation (see the icons and warnings on page S38 of the Student Book). Have materials prepared to take outdoors.

1. Once you are outdoors, set up the experiment in a suitable location. Encourage students to observe what the soil looks like before the experiment begins. Make sure that students prepare to record observations. Once the fan is turned on, change may occur rapidly.

Investigating Soil

INVESTIGATING SOIL

⚠ Blowing dust can cause eye irritation. Stay away from where the soil is blowing. Use eye and mouth protection.

2. Put on safety goggles and a dust mask.

   Turn the fan on to a low speed. If the fan is far enough away from the soil, none of the soil particles will be moved by the fan.

   Increase the speed of the fan to medium or high speed. If some of the soil particles are moved, observe carefully how they are moved.

   a) Record your observations.

3. Turn the fan off, and slide it a bit closer to the soil.

   Turn the fan on again, and observe the movement of the soil.

   Keep moving the fan closer to the soil to model a stronger and stronger wind.

   a) Carefully, with as much detail as possible, record your observations.

4. Use your observations to answer the following questions:

   a) How does the motion of the soil particles in the wind differ from the motion of the soil particles in the water flow? How is it similar?

   b) How does the movement of the coarser soil particles differ from the movement of the finer soil particles?

   c) Are some of the soil particles too large or too heavy to be moved by even a strong flow of air?

   d) How far do you think the finest soil particles might travel in the wind?

5. In your small groups discuss further investigations you would like to do based on your exploration. For example, you may study how compacting the soil, or watering the soil might affect the amount of erosion by the wind.

   Share your ideas with the class.

   Design an investigation to test if your predictions were correct.

   a) Record materials you will need, the steps of the investigation, and your prediction. Don't forget to give a reason for your prediction.

   ⚠ Have your plan approved by your teacher before you begin your investigation.

   b) With the approval of your teacher carry out your investigation. Record your findings. Were your predictions correct?

S 38
Investigating Earth Systems

202 Investigating Earth Systems

Teacher Commentary

2. Ensure that students are wearing appropriate safety equipment prior to turning on the fan. Observe students as they investigate.

   a) Encourage students to make quality journal entries. Check their journals to make sure they include dates for each entry, headings, and clear, thoughtful answers. Try to find something good about every student's journal, as well as something he/she needs to improve.

3. Be sure to turn off the fan before moving it closer to the soil. Check to make sure that students are recording their observations. They may wish to make a measurement of how far the soil moved and how far away the fan was from the soil.

   a) Students should note that more soil was moved when the fan was closer to the soil.

4. a) Soil movement will increase with increasing force of the air, as it did with as powerful an erosional agent as water.

   b) Three acceptable responses include:
   - Finer soil particles will move before coarser particles begin to move.
   - Greater quantities of fine particles will be moved than coarser particles.
   - Finer particles will move greater distances than coarser particles.

   c) Yes, some of the larger soil particles will not move, even with a strong blast of air.

   d) Answers will vary. Accept all reasonable responses. Fine particles (dust) lifted high in the atmosphere have been known to travel thousands of kilometers.

5. Encourage students to discuss their ideas and to record their questions and ideas. Consider holding a class discussion about students' questions for inquiry. Before students begin to conduct new investigations, check to see that they have a testable question.

Teaching Tip
Making Predictions
This is another opportunity for students to apply what they already know to form a prediction. Again, it is important that students give clear reasons for their predictions to consider the predictions in the light of evidence. Where student predictions do not work out, ask those students to consider the reasons they gave, and revise them based on the evidence of their observations.

a) Insist that students' records have a mechanism for evaluating the results of their experiment. How will they measure soil erosion? Examples of predictions that students might make include:
- Compacted soil will erode more slowly than loose soil because it will be harder for the wind to pick up particles of soil.
- Wet soil will erode less rapidly by wind than dry soil because the water will bind the soil together.
- Less soil will erode behind a large rock than in exposed areas because the rock will reduce or block the wind.

b) Observe students as they investigate. Make sure that students record and explain their findings.

Making Connections ...with the local area

As an extension to the activity instruct students to detect, describe, and photograph or videotape examples of soil erosion in the community.

Teacher Commentary

NOTES

Investigation 6: Soil Erosion

Digging Deeper

Soil Erosion

Water or air that moves over a soil surface applies forces to the soil particles on the surface. If the forces are large enough, they move the particles. The stronger the current or wind, the more particles are put into motion. Larger and heavier particles tend to roll or hop near the soil surface. Finer and lighter particles are carried upward from the soil surface. In nature, the finest particles may be carried for hundreds or even thousands of kilometers high in the atmosphere before they fall out!

Soil erosion is a serious problem in many areas of the world. Soil takes thousands of years to form, but much of it can be eroded by just a few unusually heavy rainstorms or strong winds.

Bare soil surfaces are very likely to be eroded by a sudden heavy rainstorm. The running water can cut a channel, called a gully, in the soil surface. Once the gully is cut, the force of the water is focused there. This deepens the channel.

Teacher Commentary

Digging Deeper

The text in the **Digging Deeper** reading section explains the basic nature of soil erosion by wind and water.

As You Read...
[As You Read questions are missing from the textbook.]
1. What are the two ways that soil is eroded?
   Soil is eroded by wind and by water.
2. What two ways can soil erosion be reduced?
   Soil erosion can be reduced by planting vegetation and by planting in rows that follow the contour of a slope.

About the Photo

The photograph on page S39 provides a real example of what students modeled in Part 2 of the investigation. Check for understanding by asking students to explain what they see and why it is occurring. The photograph shows soil erosion by wind in a region with little vegetation to hold the soil down, and dry soil, which is more easily eroded by wind than is wet soil.

Assessment Opportunity

On the board, write down the definitions for renewable resource and non-renewable resource. Ask students to read the two statements below. Instruct them to select and complete the sentence that they agree with based upon their understanding of soil and the meaning of resources as you have defined.

A. I think that soil is a renewable resource because _____.
B. I think that soil is a non-renewable resource because _____.

For definitions of renewable resources that are based upon the assumption of replenishing the resource within a human lifespan (80 years), soil is readily classified as a non-renewable resource. Students' answers should reveal whether or not they understand the difference between the times scales over which soil forms versus that which soil can be eroded. Some students may consider soil a renewable resource because it "never goes away" – it is always transported somewhere. Such responses indicate a sophisticated understanding of the conservation of matter and of the recycling of matter within the Earth system.

Investigating Soil

INVESTIGATING SOIL

One very good way to reduce soil erosion is to keep the soil surface covered with vegetation. Another way is to plant crops in rows that follow the contours of the land surface rather than running up and down a slope.

Teacher Commentary

Digging Deeper

About the Photo

The photograph at the top of page S40 depicts the formation of gullies in soil (which is explained in the text at the bottom of page S39).

The lower photograph shows how terraces on steep slopes minimize soil erosion. Terracing the slope parallel to the contours of the land surface reduces the steepness of the slope and makes the land less susceptible to gullying. This is an excellent example of conserving a natural resource (as well as a practical way for people in mountainous regions to raise crops for food and for livestock).

For additional teaching suggestions on how to present the content of this **Digging Deeper** reading section to your students, please refer to the *IES* web site.

Investigation 6: Soil Erosion

Review and Reflect

Review

1. Look again at the key question. Has the investigation helped you answer the question? Based on your investigation, how is soil worn away?
2. Suppose you were to conduct this investigation on soil that contained a lot of gravel and soil that contained fine sand. Which would you predict would erode more quickly?
3. Based on the result of your investigation and your reading, explain why soil erosion is important.

Reflect

4. Where do you think the soil that is eroded goes?
5. How could you use what you found out in this investigation in a practical way?
6. Based on what you now know, what further questions about soil erosion would be useful to investigate?

Thinking about the Earth System

7. How is water (hydrosphere) involved in soil erosion?
8. How is air (atmosphere) involved in soil erosion?
9. How could the biosphere be involved in soil erosion?
10. How does soil erosion connect to the geosphere, hydrosphere, atmosphere, and biosphere? Include any new connections on your *Earth System Connection* sheet for soil.

Thinking about Scientific Inquiry

11. How did you use models to help you study soil erosion?
12. Could you observe soil erosion without using models? If so, where would you go to make your observations?
13. What are the advantages and disadvantages of using models?

Teacher Commentary

Review and Reflect

Review

1. Yes. Soil can be worn away by wind and by water.

2. Fine sand would erode more quickly. Sand is smaller than gravel, and smaller particles are more easily eroded than large particles.

3. Farmers value soil. Losing topsoil means losing the ability to grow good crops.

Reflect

4. Eroded soil is moved downstream and downwind. The finest material is carried for long distances in suspension (carried above the bed of the land in the fluid – either air or water). Sand is likely to move for a shorter distance, and it accumulates in the form of sand dunes or sandbars in rivers.

5. Answers will vary. Prevent soil erosion by planting vegetation, reduce erosion along rivers, etc.

6. Answers will vary. How long does it take for water to erode soil? How long does it take for wind to erode soil? What determines where a gully will form?

Thinking about the Earth System

7. Flowing water causes erosion; damp soils are much less susceptible to wind erosion than very dry soils.

8. Wind causes erosion.

9. Plants prevent erosion, both by anchoring the soil and breaking the force of the wind.

10. The type of soil present affects the extent of soil erosion. Small particles travel farther than large particles. Plants keep soil anchored, preventing erosion. Wind in the atmosphere and water in the hydrosphere are the two biggest causes of soil erosion.

Thinking about Scientific Inquiry

11. The stream table modeled soil erosion by water, and the fan modeled soil erosion by wind.

12. Yes. Answers will vary. Streams, rivers, the beach, a windstorm in the desert, etc.

13. Models are convenient but they only partially represent reality.

Teacher Review

Use this section to reflect on and review the investigation. Keep in mind that your notes here are likely to be especially helpful when you teach this investigation again. Questions listed here are examples only.

Student Achievement

What evidence do you have that all students have met the science content objectives?
_____

Are there any students who need more help in reaching these objectives? If so, how can you provide this?_____
_____

What evidence do you have that all students have demonstrated their understanding of the inquiry processes?_____
_____

Which of these inquiry objectives do your students need to improve upon in future investigations? _____
_____

What evidence do the journal entries contain about what your students learned from this investigation? _____
_____

Planning

How well did this investigation fit into your class time?_____
_____

What changes can you make to improve your planning next time? _____
_____

Guiding and Facilitating Learning

How well did you focus and support inquiry while interacting with students?
_____

What changes can you make to improve classroom management for the next investigation or the next time you teach this investigation? _____
_____

Teacher Commentary

How successful were you in encouraging all students to participate fully in science learning? _____

_____

How did you encourage and model the skills values, and attitudes of scientific inquiry? _____

_____

How did you nurture collaboration among students? _____

_____

Materials and Resources

What challenges did you encounter obtaining or using materials and/or resources needed for the activity? _____

_____

What changes can you make to better obtain and better manage materials and resources next time? _____

_____

Student Evaluation

Describe how you evaluated student progress. What worked well? What needs to be improved? _____

_____

How will you adapt your evaluation methods for next time? _____

_____

Describe how you guided students in self-assessment. _____

_____

Self Evaluation

How would you rate your teaching of this investigation? _____

_____

What advice would you give to a colleague who is planning to teach this investigation? _____

_____

NOTES

Teacher Commentary

INVESTIGATION 7: USING SOIL DATA TO PLAN A GARDEN

Background Information

Horticultural and agricultural planning requires attention to many different considerations: soil characteristics, climate (rainfall, temperature, winds, growing season), drainage, topography, sunlight, and surroundings.

The best soils for gardening are those that have a mixture of particle sizes, from sand to clay, with a large fraction falling in the silt size range (from about 0.01 mm to about 0.1 mm). In terms of distribution of particle sizes, such soils fall into the category of loam soils. Such soils drain neither too rapidly nor too slowly, but they have substantial porosity.

Essential for good growth of most cultivated plants is humus: plant material that has been broken down by decay into fine-grained porous plant remains. Natural soils in semi-humid to humid areas usually have a surface layer (the H horizon) that is rich in humus. Cultivation mixes this humus uniformly downward, ideally to the lowest level reached by the roots of the crops or horticultural specimen plants, which can be as shallow as a decimeter or as deep as a meter or more. Usually, gardeners try to add more decaying plant matter, in the form of such things as compost, peat moss, shredded plant material, leaf mold, or decayed animal manure. In a certain sense, the soil in a garden is much more an "outdoor growing medium" than a natural soil! The value of this humus and other plant material is partly its supply of nutrients but even more importantly its role as a medium for strong root development and a retainer of soil moisture. Addition of plant material is beneficial for all soils, very sandy soils and very clayey soils as well as loam soils.

Plant nutrients (mainly potassium ions, phosphate ions, and soluble compounds of nitrogen, but also many trace elements that are needed for plant growth) are supplied in part by the natural constituents of the soil, but for good growth of most cultivated plants, additional nutrients need to be supplied. One school of thought advocates only use of natural soil additives like humus and other plant material, as well as various other natural fertilizers. Proponents of this approach tend to call themselves organic gardeners. According to another school of thought, the better way to supply needed plant nutrients in sufficient quantities is to use inorganic or manufactured fertilizers, which are commercially available.

Cultivated plants vary widely in their preference for, or tolerance of, soil acidity. The acidity or alkalinity of the soil, as characterized by the pH, depends on rainfall as well as soil composition. Natural rainfall is acidic, even leaving aside the "acid rain" effect caused by human activities, so in areas of abundant rainfall the soil pH is likely to be rather acidic. As water resides in the soil after a rain, its pH might be increased toward a neutral or even alkaline value, by reaction with various kinds of minerals. Most important in this respect are the carbonate minerals calcite and dolomite, the common constituents of limestones and dolostones. In such areas, dissolution of some of the carbonate minerals by the initially acidic soil water raises the pH considerably, as hydrogen ions are consumed. Soils formed on parent materials rich in carbonate minerals are said to be "sweet" and are highly desirable for most vegetable crops. Many gardeners make regular applications of ground or pulverized limestone or dolostone, widely available commercially, to raise soil pH. On the other hand, certain plants are acid-loving, and thrive only in soils with low pH. Among these are most evergreen trees and shrubs. In some areas, especially in the central and western US, application of artificial soil acidifiers is necessary for the health of such acid-loving plants.

The aspect of climate that is most important to gardeners is the microclimate of the area: what are the local conditions of temperature, precipitation, and winds very near the ground, where the plants are to grow? This depends in part upon the overall climate, but also in very important ways on the specifics of the local area. For example, the difference in nighttime low temperatures and the occurrence of late spring frosts and early fall frosts depends very much on whether the plot is on locally high ground or locally low ground. Differences of a meter in elevation over distances of tens of meters can make a major difference in temperature microclimate. Winds near the ground can be weakened by various kinds of windbreaks, natural and artificial. This is especially important for protection against prevailing cold winter winds. In regions with abundant snowfall, garden plots where snowdrifts tend to accumulate are places to be avoided. Sheltered, south-facing areas, for example, against the sunny sides of buildings, can make a difference equivalent to many tens of miles north–south in plant hardiness zones. Low-lying areas can be problem sites not just in terms of temperature but also in terms of drainage after heavy rains; most plants cannot tolerate saturated soil around their roots for extended periods of time.

Plants vary greatly in their requirements for sunlight: most, especially vegetable crops, prefer or need full sun for most of the day, but some flowers and shrubs do best in partial shade or even full shade. The garden planner needs to think carefully about sun position and the shadows cast by trees and buildings throughout the growing season. The position and duration of shade is likely to vary greatly from late spring, not long before the summer solstice (on or about 21 June), to mid-fall, well past the fall equinox (on or about 21 September.

The *Investigating Earth Systems* www.agiweb.org/ies web site also contains a variety of links to web sites that will help you deepen your understanding of content and prepare you to teach this investigation.

Teacher Commentary

Investigation Overview

In this final investigation, students will apply all of the knowledge and skills they have developed throughout the module to plan a garden for the school grounds or the local area. Students analyze the conditions of the garden site, conduct research, and interview gardeners to identify suitable plants, discuss and present their results with others, and revise and finalize their plans.

Goals and Objectives

The purpose of this activity is to review science content and **Inquiry Processes** that have been used throughout the module. It can be used as a final assessment, review for a final test, or both. As a result of this investigation, students will develop a better understanding of soil in their local area, and will improve their ability to solve a problem and effectively communicate scientific information to others.

Science Content Objectives

Students will collect evidence that:
1. Plants depend on the water content in soil to survive.
2. Different plants require different amounts of soil drainage to thrive.
3. Plants are affected by the nutrients in soil.
4. Soil type varies with location.
5. Soil nutrients can be altered through biological and chemical means.
6. Soil is part of the Earth's Systems.

Inquiry Process Skills

Students will:
1. Use sampling and testing skills to analyze soil from a local area.
2. Compare soil data to resource information about plant needs.
3. Devise methods to alter growing conditions for plants, if necessary.
4. Devise a garden plan based upon soil test results, needs of different plants, costs, and available methods of altering growing conditions.
5. Create a scale diagram of the garden plan.
6. Communicate their plan to others.
7. Arrive at a consensus on the most effective garden plan.
8. Create a presentation of the garden plan for the school administration.
9. Present the garden plan.
10. Carry out the garden plan.
11. Analyze the garden plan.

Connections to Standards and Benchmarks

In this investigation, students will discover some of the directly observable components of soil. These observations will start them on the road to understanding the National Science Education Standards and American Association for the Advancement of Science Benchmarks shown below.

Investigating Soil – Investigation 7

NSES Links
- Soil consists of weathered rocks and decomposed organic material from dead plants, animals, and bacteria. Soils are often found in layers, with each having a different chemical composition and texture.
- All organisms must be able to obtain and use resources, grow, reproduce, and maintain stable internal conditions while living in a constantly changing external environment.
- Water evaporates from the Earth's surface, rises and cools as it moves to higher elevations, condenses as rain or snow, and falls to the surface where it collects in lakes, oceans, soil, and in rocks under the ground.
- Water is a solvent. As it passes through the water cycle it dissolves minerals and gases and carries them to the oceans.

AAAS Links
- Organize information in simple tables and graphs and identify relationships they reveal.
- In any particular environment, the growth and survival of organisms depend on the physical conditions.
- Although weathered rock is the basic component of soil, the composition and texture of soil and its fertility and resistance to erosion are greatly influenced by plant roots and debris, bacteria, fungi, worms, insects, rodents, and other organisms.

Teacher Commentary

Preparation and Materials Needed

Preparation

Implementing the plan in your school grounds will depend upon factors such as administration permission and an available spot. The time of year will also be important. Garden options your students could consider might be local parks, vacant lots, senior centers, or other suitable places. It will be important to get permission from those in authority over the garden site before making any plans to plant.

If your students live in an urban setting with no land space for a small garden, you can substitute a window box or small box garden. Another alternative would be to construct mini-gardens in a large-size aluminum roasting pan. This will work if your students choose small plants with shallow root systems.

In a gardening magazine, find the addresses for companies that offer free catalogs. Make these addresses available for students so they can send away for a gardening catalog. Information about which plants grow well in which soils and environments is available in gardening and horticultural books. A nursery and/or garden supply store will also be able to help with advice.

Materials

For this investigation each group will decide what materials are needed.

Investigating Soil

INVESTIGATING SOIL

Investigation 7:

Using Soil Data to Plan a Garden

Putting It All Together

Materials Needed

For this investigation your group will decide what materials are needed.

Key Question

Before you begin, first think about this key question.

What soil mixtures are best for plants?

In this last investigation you will have a chance to apply all that you have learned about soil to solve a practical problem. You are going to plan a garden for your school grounds, or your local area. Think about all the things that you have learned about soil that you will need to answer the key question.
Share your ideas with others in your class.

Explore Questions

⚠ Consider all necessary safety precautions at every phase of planning and carrying out the project.

Investigate

1. Choose an area on your school grounds or in your community where you think you might want to plant a garden.

    Obtain permission to use the site for planning the garden.

S 42
Investigating Earth Systems

Teacher Commentary

Key Question

Have students respond to the **Key Question**, "What soil mixtures are best for plants?" You may want to write the question on the board. Help students to feel comfortable defining "soil mixture" in their own terms. Prompt them to think about what they have learned. What are examples of soil mixtures? What have you learned about the materials in soil in previous investigations that helps you to think about what makes them suitable for certain types of plants? Give students five minutes to respond in their journals. Discuss students' ideas. Require that students explain their thinking.

Student Conceptions

Students should come to the consensus that different plants grow in different environments.

> ### Assessing the Final Investigation
> Students' work throughout the module culminates with this final investigation. To complete it, students need a working knowledge of previous activities. Because it refers back to the previous steps, the last investigation is a good review and a chance to demonstrate proficiency. For an idea how to use the last investigation as a performance-based exam, see the section in the **Appendix**. Make sure that you spend time with students reviewing the criteria upon which their final project will be assessed.

Investigate
Teaching Suggestions and Sample Answers

1. Decide on the composition of student groups. Because this is the last investigation of the module, you may want to use it as an assessment tool or as a review for a final test (or both). Choose groups carefully so that there is a mix of abilities and good group dynamics. If the group dynamics are not good, change them before the end of the first part of this investigation.

   If you intend for students to plan a garden on the school grounds, make sure permission has been granted to use the land and that it is in a safe location.

Investigation 7: Using Soil Data to Plan a Garden

2. When you have permission to use the site for planning, obtain some soil samples from the area.

3. Divide the work of analyzing the samples among the different groups in the class. Look over all the previous investigations you have done. Decide how your group will analyze the samples. Review the method you will use.

   a) In your notebook list the materials you will need. Record the method you will use for your analysis.

4. Run your tests.

   a) Record all your data.

   b) As a class, combine your results. You may wish to use a data table similar to the one shown. Change it to suit your needs. It may be helpful to use a separate table for each sample.

Soil Analysis Data Table for Soil Sample			
Soil Analysis	Test 1	Test 2	Test 3
Observational data			
Soil components			
Water content			
Water drainage			
pH level			
Phosphorus level			
Nitrogen level			
Potassium level			
Other aspects			

Teacher Commentary

2. Stress to your students that the way they place particular plants in particular locations must be supported by their research data on soil. Review what you mean by soil analysis data before your students begin. Students will then need to collect soil samples from the garden site. Review safety procedures to ensure that anything students do in this final investigation is safe.

   Students may do up to three tests on the soil. Advise students to think about what kind of tests they will run before collecting the soil (the quantity of soil collected might depend on its intended purpose). You will need to have space available to store students' soil samples and any materials they will be using.

   ### Making Connections ...with Mathematics

   There are many opportunities for students to use mathematics in this investigation. They will need to make measurements when planning the garden shape and size; the distribution of plants' quantities of seeds or plantings; soil chemical levels; and so on. Students' use of mathematics here also provides you with assessment data.

3. a) Advise students that they need to bring any required materials to class. Emphasize that they will be gathering evidence to answer the **Key Question**. Ask them what kind of evidence they think would be valuable.

4. Look at students' data tables. Check to make sure they have an accurate method of recording data.

   a) Observe students as they conduct their tests. Check to make sure that they are recording their data and following safe laboratory procedures.

   b) When students have completed their analyses, hold a class discussion about the results of soil analyses. Even though this investigation has a strong assessment component to it, help students to feel comfortable sharing their findings at this stage so that everyone can benefit from the data collected. This mirrors scientific inquiry processes where the research of others is available to all.

Investigating Soil

INVESTIGATING SOIL

5. Use the completed data table and any gardening resources to decide which plants will do well in your area. Consider how you will balance the following factors:
   - the soil type;
   - plants that grow well in that soil type;
   - plants that you would like to grow;
   - ways in which you could change your soil without harming the environment and without costing a lot;
   - the light available;
   - the climate in your area;
   - direction and steepness of the ground slope.

   a) Record the names of the plants you wish to grow. Note the growing conditions required for each plant.

   b) Make a detailed plan of your garden.

6. Try to arrange interviews with one or more experienced local gardeners. This is very important. You have planned carefully. However, these people may know things about your area that you may not have thought of.

   a) After listening to their advice, revise your plan if you think you need to.

Teacher Commentary

5. Students should bring gardening catalogs to class. Students can also conduct research on the Internet, either during class or as a homework assignment. There are a number of web sites on soil science that your students could access, but be advised that these are often quite sophisticated in the amount and level of information they provide. The *IES* web site has links to resources that will be helpful at www.agiweb.org/ies/.

    They may want to look at the types of plants suitable to:
    - high pH
    - low pH
    - neutral pH
    - poorly aerated soil
    - poorly drained soil
    - particular proportions of nutrients (nitrogen, potassium, phosphorus)
    - particular proportions of soil components (clay, sand, humus, etc.)

    a) Based upon their research and their analysis of the soil and the gardening site, students should record the names of the plants they wish to grow and the growing conditions required for each plant.

    b) The "detailed plan" can, and probably should, include a sketch. The sketch included on page S45 of the Student Book is to encourage them to plan their gardens carefully and support their plans with their soil data.

6. Arrange and conduct an interview with a local gardener. The interview does not have to be with a professional gardener. Family members or neighborhood residents are fine too. You might find it useful to contact gardening experts in your area and put them on "standby" for this final activity. You could also ask for parent volunteers who are knowledgeable gardeners. Have students videotape, record, or transcribe their interviews.

    a) Adjusting the plan, in the light of new information, is also an important science process. Remind students to revisit their plans after a visit from a local gardening expert.

Investigation 7: Using Soil Data to Plan a Garden

7. Share your plan with the rest of the class. Make sure that your garden plan, and all the information supporting it, is in a form that can be seen and understood by others.

   As a class, work towards one overall plan.

   Develop a presentation of your plan that you can use to convince the persons in charge to carry out your ideas. Here is an example of a plan to help you get started.

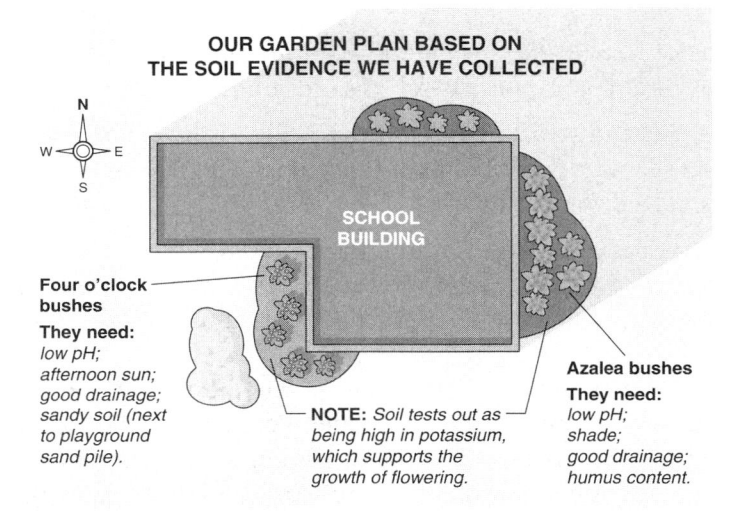

   OUR GARDEN PLAN BASED ON THE SOIL EVIDENCE WE HAVE COLLECTED

   Four o'clock bushes
   They need:
   low pH;
   afternoon sun;
   good drainage;
   sandy soil (next to playground sand pile).

   NOTE: Soil tests out as being high in potassium, which supports the growth of flowering.

   Azalea bushes
   They need:
   low pH;
   shade;
   good drainage;
   humus content.

8. To complete your plan, you will need to figure out costs. Check prices for:
   - the plants you plan to include;
   - any materials you need to add to the soil;
   - any other items you will need to carry out your plan.

   a) Record the costs needed to carry out your plan.

9. Present your plan. If possible, carry out your ideas.

Teacher Commentary

7. The importance of disseminating results has been stressed throughout the module. Students should, by now, be able to understand and use this important science process. Yet, your students may need help in presenting their garden plan and supporting data in an easily seen and understood form. You could provide them with butcher paper or poster board and markers. It helps to do things on a large scale for a clear presentation.

   Have students briefly share their information. Plans can be consolidated or each group can do its own plan. Focus on presentation skills. In Step 9, students will present their final plan. Reinforce positive aspects of presentations. Emphasize the use of visual elements.

   > **Making Connections ...with technology**
   >
   > A number of software programs are available for landscape planning. Some students may wish to make use of these programs in their final presentations.

8. When planning a garden, cost is always a factor. While your students may choose plants that would grow well in the school's soil, some of these could be cost-prohibitive. Students may be limited by their choice of garden catalogs. When unable to find the price of a particular item, estimate by using the closest related item you can find.

9. Help students to think about what they will say in the time that you have budgeted for their oral presentations the next day. You will need to budget time for each presentation and a brief question and answer session so that all groups can present their results in one class period. Be sure to discuss any criteria for assessing presentations before students begin to prepare their presentations. Help students to think about how the presentation they are constructing might become part of their information materials for the final investigation.

   Students present their final plans. Depending upon your school's resources, it might also be a good idea to ask students to make an 8" x 11" version of their plans to be photocopied and put into a class book. They can include these plans in their journals.

   > **Assessment Strategy**
   >
   > Peer review is an important part of the scientific process. Help students critique their colleagues' presentations by providing sentence stems that they should complete for their reviews. Examples include:
   > - This presentation was effective at showing...
   > - This presentation helped me to understand...
   > - This presentation needed work on...
   > - To improve this presentation, I would suggest that you...

Assessment Tool
Student Presentation Evaluation Form

The **Student Presentation Evaluation Form** provides simple guidelines for assessing presentations. Adapt and modify the evaluation form to suit your needs. If you decide to use the form, provide it to students and discuss it with them before they begin to prepare their work. Students should always know the expectations for their work before they begin to prepare for an assessment.

Teacher Commentary

NOTES

Investigating Soil

INVESTIGATING SOIL

Review and Reflect

Review

1. Were there any surprises in other groups' plans or evidence? If so, how can you explain them?
2. How was your plan the same and/or different from those of other groups?

Reflect

3. How has this investigation given you a better understanding of the relationship between soil and plants?
4. Why is the relationship between soil and plants important?
5. Why is it important to prevent soil erosion?

Thinking about the Earth System

6. How do plants and soil relate to the Earth Systems?

Thinking about Scientific Inquiry

7. Explain how you have applied the processes of scientific inquiry to solve a practical problem.

Evidence for Ideas

Teacher Commentary

Review and Reflect

Review
1. Answers will vary.

2. Answers will vary.

Reflect
3. Answers will vary

4. The biosphere depends on plants to produce food and oxygen. Land plants could not exist without soil.

5. Soil takes a very long time to produce (almost always much longer than human lifetimes), and can be eroded in a very short period of time.

Thinking about the Earth System
6. Plants need soil to grow. Soil develops, in part, from plants breaking down bedrock and from decomposing plant material.

Thinking about Scientific Inquiry
7. Observations were made, questions were framed on the basis of observations, predictions were made, an investigation was designed to test the predictions, evidence was used to accept or reject predictions, and new questions were posed.

Teacher Review

Use this section to reflect on and review the investigation. Keep in mind that your notes here are likely to be especially helpful when you teach this investigation again. Questions listed here are examples only.

Student Achievement

What evidence do you have that all students have met the science content objectives?

_____

Are there any students who need more help in reaching these objectives? If so, how can you provide this? _____

_____

What evidence do you have that all students have demonstrated their understanding of the inquiry processes? _____

_____

Which of these inquiry objectives do your students need to improve upon in future investigations? _____

_____

What evidence do the journal entries contain about what your students learned from this investigation? _____

_____

Planning

How well did this investigation fit into your class time? _____

_____

What changes can you make to improve your planning next time? _____

_____

Guiding and Facilitating Learning

How well did you focus and support inquiry while interacting with students?

_____

What changes can you make to improve classroom management for the next investigation or the next time you teach this investigation? _____

_____

Teacher Commentary

How successful were you in encouraging all students to participate fully in science learning? _____

_____

How did you encourage and model the skills values, and attitudes of scientific inquiry? _____

_____

How did you nurture collaboration among students? _____

_____

Materials and Resources

What challenges did you encounter obtaining or using materials and/or resources needed for the activity? _____

_____

What changes can you make to better obtain and better manage materials and resources next time? _____

_____

Student Evaluation

Describe how you evaluated student progress. What worked well? What needs to be improved? _____

_____

How will you adapt your evaluation methods for next time? _____

_____

Describe how you guided students in self-assessment. _____

_____

Self Evaluation

How would you rate your teaching of this investigation? _____

_____

What advice would you give to a colleague who is planning to teach this investigation? _____

_____

Investigating Soil

Reflecting

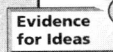

Evidence for Ideas

Back to the Beginning
You have been investigating soil in many ways. How have your ideas about soil changed from when you started? Look at the following questions and write down your ideas now.

- What is soil, and what is it made of?
- How is soil formed and how does it wear away?
- Why is soil important, and why is it important for you to know about soil?
- What questions do you have about soil?

How has your thinking about soil changed?

Thinking about the Earth System
At the end of each investigation, you thought about how your findings connected with the Earth system. Consider what you have learned about the Earth system. Refer to the *Earth System Connection* sheet that you have been building up throughout this module.

- What connections between soil and the Earth system have you been able to find?

Thinking about Scientific Inquiry
You have used inquiry processes throughout the module. Review the investigations you have done and the inquiry processes you have used.

- What scientific inquiry processes did you use?
- How did scientific inquiry processes help you learn about soil?

Not so much an ending, as a new beginning!

This investigation into soil is now completed. However, this is not the end of the story. You will see soil where you live, and everywhere you travel. Be alert for opportunities to observe soils and add to your understanding.

Reflecting

This is the point at which your students review what they have learned throughout the module. This review is very important. Allow students time to work on this in a thoughtful way.

Back to the Beginning

These four questions were used as a pre-assessment. Encourage students to complete this final review without looking at their journal entries from the beginning of the module. Their initial entries may influence their responses.

When students have completed their writing, encourage them to revisit their initial answers from the pre-assessment. Compare their writings at the end of the unit to their writings at the beginning. It is important that students are not left with the impression that they now know all there is to know about soil. Emphasize that learning is a continuous process throughout our lives.

> ### Assessment Opportunity
> Comparisons between students' initial answers to these questions (in the pre-assessment at the beginning of the module) and those they are now able to give provide valuable data for assessment.

Thinking about the Earth System

Now that your students are at the end of this module, ask them to make connections between soil and Earth's Systems. You may want to do this through a concept map. This is an opportunity for you to gauge how well students have developed their understanding of the Earth System for assessment purposes.

Thinking about Scientific Inquiry

To help students understand the relevance of these processes to their lives, ask them to think of everyday examples of when they use these processes (finding out where a misplaced book has gone; forming an opinion about a new TV show; winning an argument).

NOTES

Appendices

Investigating Soil: Alternative End-of-Module Assessment

Part A
Write the letter of the term from column B that matches the description in Column A.

Column A
1. The removal of soil by running water
2. Important nutrient in soil
3. When dropped into water it will settle faster than sand
4. Substance with a pH higher than 7
5. Test of the flow of water through soil
6. A process by which rock breaks down into soil
7. The finest solid particles in soil

Column B
A. acid
B. base
C. bedrock
D. gravel
E. weathering
F. clay
G. erosion
H. percolation
I. variable
J. nitrogen

Part B: Multiple Choice
Provide the letter of the choice that best answers each question below.

8. A topographic map shows
   A. the elevation of the land in a region.
   B. the number of people that live in a region.
   C. the quality of soil in a region.
   D. the kinds of rocks in a region.

9. As the speed of the wind increases, the amount of soil eroded
   A. decreases.
   B. increases.
   C. decreases, then increases.
   D. stays the same.

10. The main components of soil are inorganic matter and
    A. fertilizer.
    B. rock fragments.
    C. sand.
    D. plant and animal material.

11. Wind and water that move across a soil surface
    A. contaminate soil particles.
    B. do not affect soil particles
    C. apply forces to soil particles.
    D. cannot pick up soil particles.

12. One way to reduce soil erosion is to
    A. reduce ground vegetation.
    B. increase ground vegetation.
    C. plant crops in lines that run perpendicular to the contour of the land.
    D. avoid adding fertilizer to crops.

13. Flowering dogwood trees do not grow well in acidic or basic soils. Which of the following would be the best soil pH for growing a flowering dogwood?
    A. 6
    B. 7
    C. 9
    D. 10

14. The soil profile for a soil that just started to form would
    A. consist mainly of broken rock fragments.
    B. have a thick layer of organic matter.
    C. contain at least three different horizons or zones.
    D. be about 24 to 36 inches deep.

15. The thickest soil is most likely to form in regions with a
    A. cold, dry climate.
    B. cold, wet climate.
    C. warm, dry climate.
    D. warm, wet climate.

16. A core sample of an old, mature soil is most likely to
    A. consist of several layers of different kinds of material.
    B. consist of one type of material.
    C. have more organic matter deeper in the core than near the top of the core.
    D. have more rock fragments near the top of the core than deeper in the core.

17. Water will pass through soil faster if the soil
    A. is very compacted.
    B. is made up mostly of fine particles.
    C. is made up mostly of large particles.
    D. has no organic matter in it.

Investigating Soil

18. Organic matter in soil comes from
    A. bacteria and decomposed plants and animals.
    B. clay particles.
    C. the weathering of sand and gravel.
    D. chemical reactions between water and rock.

19. The formation of soil is best described as coming from interactions between
    A. the geosphere and hydrosphere.
    B. the geosphere and atmosphere.
    C. the geosphere and biosphere.
    D. the geosphere, hydrosphere, atmosphere, and biosphere.

20. The two most important factors that determine the type of soil that forms are
    A. plowing and fertilizers.
    B. bedrock and climate.
    C. elevation and moisture.
    D. land slope and elevation.

Answer Key

1. G
2. J
3. D
4. B
5. H
6. E
7. F
8. A
9. B
10. D
11. C
12. B
13. B
14. A
15. D
16. A
17. C
18. A
19. D
20. B

Investigating Earth Systems Assessment Tools

Assessing the Student *IES* Journal

- Journal–Entry Evaluation Sheet
- Journal–Entry Checklist
- Key–Question Evaluation Sheet
- Investigation Journal–Entry Evaluation Sheet
- Review and Reflect Journal–Entry Evaluation Sheet

Assisting Students with Self Evaluation

- Group–Participation Evaluation Sheet I
- Group–Participation Evaluation Sheet II

Assessing the Final Investigation

- Final Investigation Evaluation Sheet
- Student Presentation Evaluation Form

References

- Doran, R., Chan, F., and Tamir, P. (1998). *Science Educator's Guide to Assessment.*
- Leonard, W.H., and Penick, J.E. (1998). *Biology – A Community Context.* South-Western Educational Publishing. Cincinnati, Ohio.

Journal–Entry Evaluation Sheet

Name: _____ Date: _____ Module: _____

Explanation: The journal is an important component of each *IES* module. In using the journal as you investigate Earth science questions, you are mirroring what scientists do. The criteria, along with others that your teacher may add, will be used to evaluate the quality of your journal entries. Use these criteria, along with instructions within investigations, as a guide.

Criteria

1. Entry Made

      1   2   3   4   5   6   7   8   9   10   _____

   Blank   Nominal   Above average Thorough

2. Detail

      1   2   3   4   5   6   7   8   9   10   _____

   Few dates   Half the time   Most days   Daily
   Little detail   Some detail   Good detail   Excellent detail

3. Clarity

      1   2   3   4   5   6   7   8   9   10   _____

   Vague   Becoming clearer   Clearly expressed
   Disorganized   well organized

4. Data Collection/Analysis

      1   2   3   4   5   6   7   8   9   10   _____

   Data collected   Data collected,   Data collected
   Not analyzed   some analyzed   and analyzed

5. Originality

      1   2   3   4   5   6   7   8   9   10   _____

   Little evidence   Some evidence   Strong evidence
   of originality   of originality   of originality

6. Reasoning/Higher-Order Thinking

      1   2   3   4   5   6   7   8   9   10   _____

   Little evidence   Some evidence   Strong evidence
   of thoughtfulness of thoughtfulness of thoughtfulness

7. Other

      1   2   3   4   5   6   7   8   9   10   _____

8. Other

      1   2   3   4   5   6   7   8   9   10   _____

Investigating Soil

Journal–Entry Checklist

Name: _____ Date: _____ Module: _____

Explanation: The journal is an important component of each *IES* module. In using the journal as you investigate Earth science questions, you are mirroring what scientists do. The criteria, along with others that your teacher may add, will be used to evaluate the quality of your journal entries. Use these criteria, along with instructions within investigations, as a guide.

Criteria

1. Makes entries   _____

2. Provides dates and details   _____

3. Entry is clear and organized   _____

4. Shows data collected   _____

5. Analyzes data collected   _____

6. Shows originality in presentation   _____

7. Shows evidence of higher-order thinking   _____

8. Other   _____

9. Other   _____

Total Earned   _____

Total Possible   _____

Comments:

Key–Question Evaluation Sheet

Name: _____ Date: _____ Module: _____

Shows evidence of prior knowledge
   No Entry   Fair   Strong
   0   1   2   3   4

Reflects discussion with classmates
   No Entry   Fair   Strong
   0   1   2   3   4

Additional Comments

Key–Question Evaluation Sheet

Name: _____ Date: _____ Module: _____

Shows evidence of prior knowledge
   No Entry   Fair   Strong
   0   1   2   3   4

Reflects discussion with classmates
   No Entry   Fair   Strong
   0   1   2   3   4

Additional Comments

Key–Question Evaluation Sheet

Name: _____ Date: _____ Module: _____

Shows evidence of prior knowledge
   No Entry   Fair   Strong
   0   1   2   3   4

Reflects discussion with classmates
   No Entry   Fair   Strong
   0   1   2   3   4

Additional Comments

Investigation Journal–Entry Evaluation Sheet

Name: _____ Date: _____ Module: _____

Criteria

1. Completeness of written investigation
       1   2   3   4   5   6   7   8   9   10   _____
   Blank   Incomplete   Thorough

2. Participation in investigations
       1   2   3   4   5   6   7   8   9   10   _____
   None or little;   Needs minimal guidance,   Leads, is inquisitive,
   unable to guide   sometimes helping others   persistent, focused
   self

3. Skills attained
       1   2   3   4   5   6   7   8   9   10   _____
   Few skills   Tends to use some   High degree of
   evident   appropriate skills   appropriate skills used

4. Investigation Design
       1   2   3   4   5   6   7   8   9   10   _____
   Variables not   Sometimes   Considers variables
   considered   considers variables, Sound rationale for
   techniques   uses logical techniques techniques
   illogical

5. Conceptual understanding of content
       1   2   3   4   5   6   7   8   9   10   _____
   No evidence   Approaches understanding Exceeds expectations
   of understanding of most concepts   for content attainment

6. Ability to explain/discuss inquiry
       1   2   3   4   5   6   7   8   9   10   _____
   Unable to   Some ability to   Uses scientific reasoning
   articulate   explain/discuss   to explain any
   scientific thought the inquiry   aspect of the inquiry

7. Other
       1   2   3   4   5   6   7   8   9   10   _____

8. Other
       1   2   3   4   5   6   7   8   9   10   _____

Review and Reflect Journal–Entry Evaluation Sheet

Name: _____ Date: _____ Module: _____

Criteria	Blank		Fair		Excellent	
Thoroughness of answers	0	1	2	3	4	5
Content of answers	0	1	2	3	4	5
Other	0	1	2	3	4	5

Review and Reflect Journal–Entry Evaluation Sheet

Name: _____ Date: _____ Module: _____

Criteria	Blank		Fair		Excellent	
Thoroughness of answers	0	1	2	3	4	5
Content of answers	0	1	2	3	4	5
Other	0	1	2	3	4	5

Review and Reflect Journal–Entry Evaluation Sheet

Name: _____ Date: _____ Module: _____

Criteria	Blank		Fair		Excellent	
Thoroughness of answers	0	1	2	3	4	5
Content of answers	0	1	2	3	4	5
Other	0	1	2	3	4	5

Group–Participation Evaluation Sheet I

Key:
4 = Worked on his/her part and assisted others
3 = Worked on his/her part
2 = Worked on part less than half the time
1 = Interfered with the work of others
0 = No work

My name is _____ .   I give myself a _____

The other people in my group are: I give each person:

A. _____   _____

B. _____   _____

C. _____   _____

D. _____   _____

Key:
4 = Worked on his/her part and assisted others
3 = Worked on his/her part
2 = Worked on part less than half the time
1 = Interfered with the work of others
0 = No work

My name is _____ .

The other people in my group are:

A. _____

B. _____

C. _____

D. _____ .

Group–Participation Evaluation Sheet II

Name: _____ Date: _____ Module: _____

Key:
Highest rating _____
Lowest rating _____

1. In the chart, rate each person in your group, including yourself.

	Names of Group Members				
Quality of Work					
Quantity of Work					
Cooperativeness					
Other Comments					

2. What went well in your investigation?

3. If you could repeat the investigation, how would you change it?

Final Investigation Evaluation Sheet

Alerting students

Before your students begin the final investigation, they must understand what is expected of them and how they will be evaluated on their performance. Review the task thoroughly, setting time guidelines and parameters (whom they may work with, what materials they can use, etc.). Spell out the evaluation criteria for each level of proficiency shown below. Use three categories for a 3-point scale (Achieved, Approaching, Attempting). If you prefer a 5-point scale, add the final two categories.

Name: _____ Date: _____ Module: _____

	Understanding of concepts and inquiry	Use of evidence to explain and support results	Communication of ideas	Thoroughness of work
Exceeding proficiency **5**	Demonstrates complete and unambiguous understanding of the problem and inquiry processes used.	Uses all evidence from inquiry that is factually relevant, accurate, and consistent with explanations offered.	Communicates ideas clearly and in a compelling and elegant manner to the intended audience.	Goes beyond all deliverables agreed upon for the project and has extended the data collection and analysis.
Achieved proficiency **4**	Demonstrates fairly complete and reasonably clear understanding of the problem and inquiry processes used.	Uses the major evidence from inquiry that is relevant and consistent with explanations offered.	Communicates ideas clearly and coherently to the intended audience.	Includes all of the deliverables agreed upon for the project.
Approaching proficiency **3**	Demonstrates general, yet somewhat limited understanding of the problem and inquiry processes used.	Uses evidence from inquiry to support explanations but may mix fact with opinion, omit significant evidence, or use evidence that is not totally accurate.	Completes the task satisfactorily but communication of ideas is incomplete, muddled, or unclear.	Work largely complete but missing one of the deliverables agreed upon for the project.
Attempting proficiency **2**	Demonstrates only a very general understanding of the problem and inquiry processes used.	Uses generalities or opinion more than evidence from inquiry to support explanations.	Communication of ideas is difficult to understand or unclear.	Work missing several of the deliverables agreed upon for the project.
Non-proficient **1**	Demonstrates vague or little understanding of the problem and inquiry processes used.	Uses limited evidence to support explanations or does not attempt to support explanations.	Communication of ideas is brief, vague, and/or not understandable.	Work largely incomplete; missing many of the deliverables agreed upon for the project.

Student Presentation Evaluation Form

Student Name_____ Date_____

Topic_____

	Excellent		Fair		Poor
Quality of ideas	4	3	2		1
Ability to answer questions	4	3	2		1
Overall comprehension	4	3	2		1

COMMENTS _____
_____
_____
_____
_____

Student Presentation Evaluation Form

Student Name_____ Date_____

Topic_____

	Excellent		Fair		Poor
Quality of ideas	4	3	2		1
Ability to answer questions	4	3	2		1
Overall comprehension	4	3	2		1

COMMENTS _____
_____
_____
_____
_____

Blackline Master *Soil* P.1

Questions about Soil

- What is soil and what is it made of?

- How is soil formed and how does it wear away?

- Why is soil important and why is it important for you to know about soil?

- What questions do you have about soil?

Use with *Soil* Pre-assessment.

Blackline Master *Soil* P.2

Student Journal
Investigating Soil

Name: _____

Group Members:

1. _____

2. _____

3. _____

4. _____

Teacher: _____

Class: _____

Dates of Investigation:

Start _____ Complete _____

Keep this journal with you at all times during your study of
Investigating Soil.

Use with *Soil* Pre-assessment.

Investigating Soil **253**

Blackline Master *Soil* I.1

Name: _____

Earth System Connections   Soil

When you finish an investigation, use this sheet to record any links you can make with the Earth system. By the end of the module you should have as complete a diagram as possible.

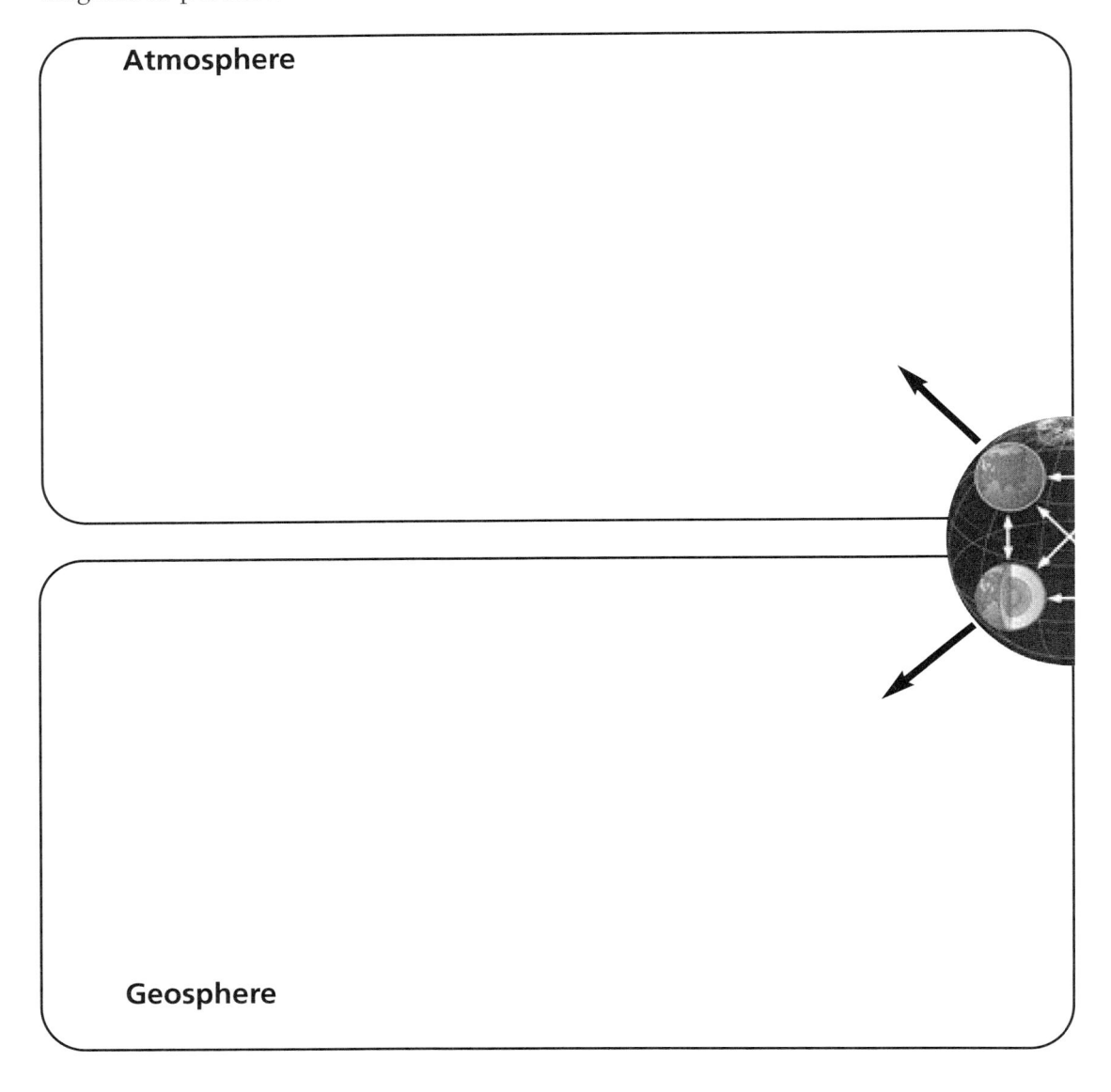

Atmosphere

Geosphere

Use with *Soil* Introducing the Earth System.

Biosphere

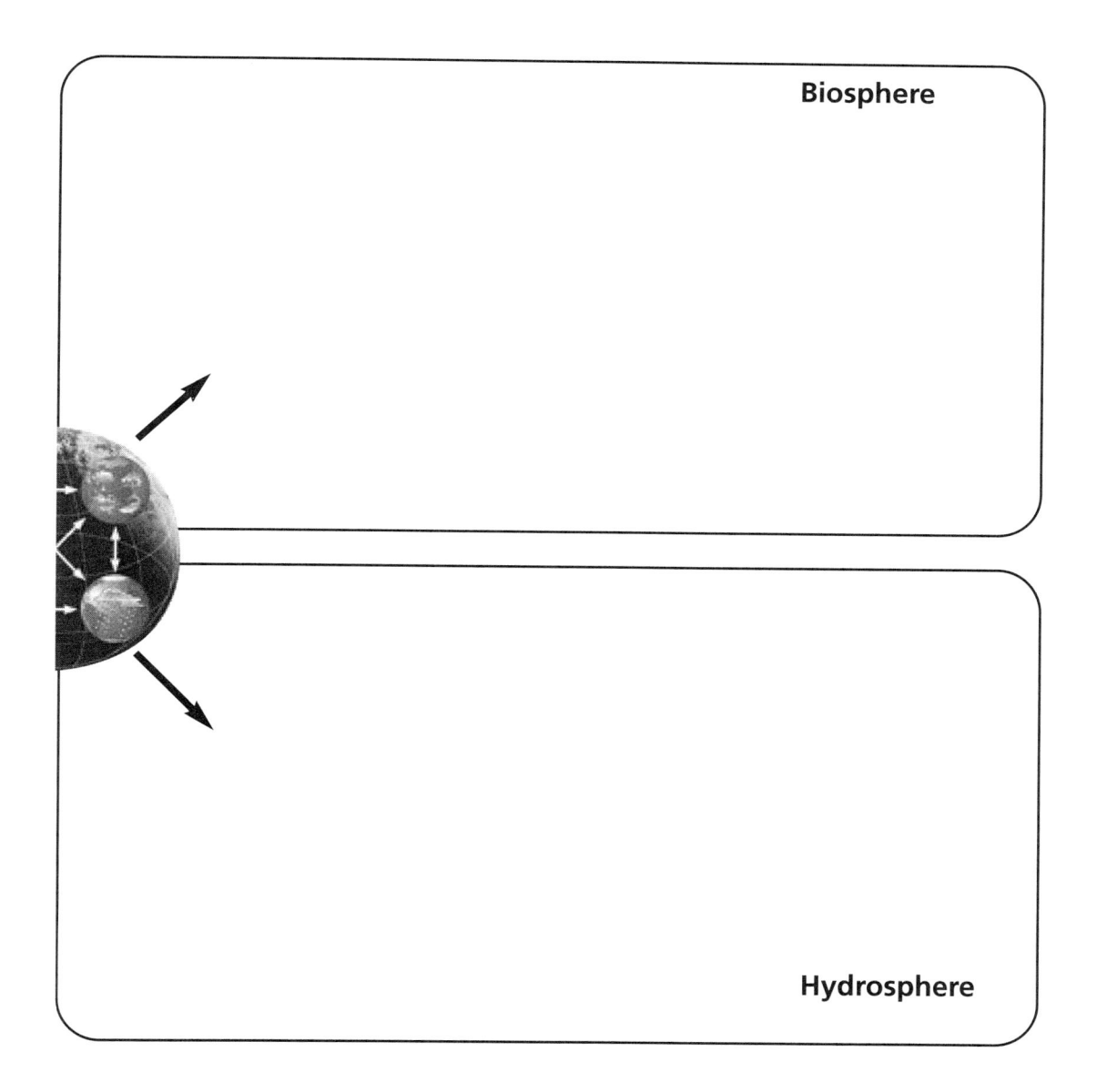

Hydrosphere

Blackline Master *Soil* I.2

Inquiry Processes

- Explore questions to answer by inquiry.
- Design an investigation. 
- Conduct an investigation. 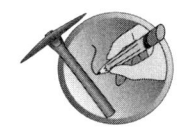
- Collect and review data using tools. 
- Use evidence to develop ideas. 
- Consider evidence for explanations. 
- Seek alternative explanations. 
- Show evidence and reasons to others. 
- Use mathematics for science inquiry. 

Use with *Soil* Introducing the Earth System.

Blackline Master *Soil* 1.1

Soil Data

Name: _____

Seeing data

_____

_____

_____

_____

Touching data

_____

_____

_____

_____

Other data

_____

_____

_____

_____

Smelling data

_____

_____

_____

_____

Use with *Soil* Investigation 1: Beginning to Investigate Soil.

Investigating Soil 257

Blackline Master *Soil* **3.1**

Name: _____

Measuring and Calculating Mass of Materials Found in Soil

1. Measure the mass of the entire dry sample.
Mass of entire dry sample: _____

2. Measure the mass of each piece of poster board.
Mass of poster board:

1 _____   2 _____

3 _____   4 _____

3. Separate the soil into components. Follow Steps 1 to 6 on pages S12 – S14 in your *Investigating Soil* textbook.

4. Measure the mass of each sample (plus poster board).
Mass of each sample and poster board:

1 _____   2 _____

3 _____   4 _____

5. Subtract the mass of the poster board from the mass of poster board plus sample. This gives you the mass of each sample.
Mass of each sample:

1 _____   2 _____

3 _____   4 _____

6. Divide the mass of each sample by the total mass of the sample and multiply the result by 100. This gives you the relative percent of each component by mass.

Mass of each sample (results from Step 5) × 100%
Mass of entire dry sample (result from Step 1):

Relative percent of each component by mass:

1 _____   2 _____

3 _____   4 _____

Use with *Soil* Investigation 3: Separating Soil by Sieving.

Blackline Master *Soil* **4.1**

Name: _____

Data from Analysis of Soil-Core Sample

Material group	Type of material(s)	Color(s) of materials	Particle size	Name of material
1				
2				
3				
4				
5				

Use with *Soil* Investigation 4: Examining Core Samples of Soil.

Blackline Master *Soil* 5.1

Soil Profile

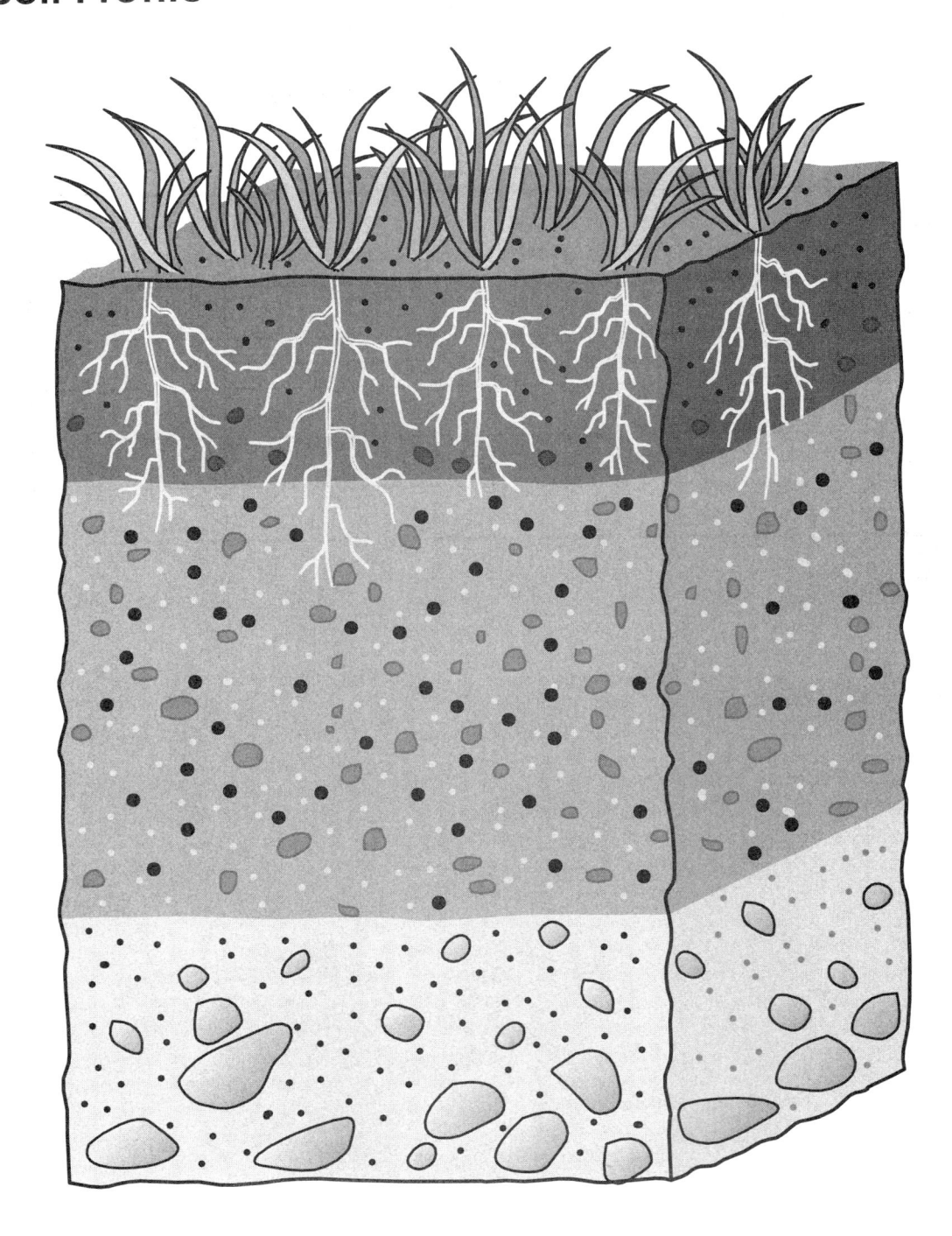

Use with *Soil* Investigation 5: Water and Other Chemicals in Soil.

Blackline Master *Soil* 6.1

Stream Table Setup

Use with *Soil* Investigation 6: Soil Erosion.

NOTES